THE VETIVER SYSTEM FOR IMPROVING WATER QUALITY

PREVENTION AND TREATMENT OF CONTAMINATED WATER AND LAND

Second Edition (2015)

By

Paul Truong and Luu Thai Danh

Second Edition 2015
First Edition 2008
Published by The Vetiver Network International

Cover photo – Boonah (Queensland, Australia) sewage effluent treatment. A Vetiver System phytoremediation application

PREFACE

The Vetiver System (VS) is dependent on the use of a very unique tropical plant, Vetiver grass, *Chrysopogon zizanioides*. The plant can be grown over a very wide range of climatic and soil conditions, and if planted correctly can be used virtually anywhere under tropical, semi-tropical, and Mediterranean climates. It has characteristics that in totality are unique to a single species. When Vetiver grass is grown (generally) in the form of a narrow self-sustaining hedgerow it exhibits special characteristics that are essential to many of the different applications that comprise the Vetiver System. Vetiver grass can be used for applications that will protect river basins and watersheds against environmental damage, particularly point source environmental problems relating to: (1) sediment flows, and (2) excess nutrients, heavy metals and pesticides in leachate from toxic sources. The two are closely linked.

This handbook, originally published in 2008 has been completely revised to reflect the significant gains in knowledge, both research and field experience that has occurred in most parts of the tropics and semi-tropics over the past seven years. Over that time Vetiver System applications for the decontamination or containment of polluted water and land has shown significant success, and as a result there is a growing interest in its application for the mitigation of many water/land related problems that are nowadays often overwhelming communities, both rural and urban, due to poverty, population pressures, lack of public financing, and climate change. The applications described in this handbook can be used at various scales by a very wide range of users, and have particular potential for providing the means for communities to create resilience, at low cost (often independent from formal public services), that will allow a more secure and better quality of life.

An important objective of this handbook is to introduce the Vetiver System to planners, design engineers and other potential users, who continue to be unaware of the effectiveness of the Vetiver System for improving water quality, particularly that associated with effluent discharge and leachate flows from industrial sites, contaminated mining sites, polluted urban/domestic waste water, and sadly and more too often agro-chemical contaminated agricultural land.

This handbook has been revised by Paul Truong and Luu Thai Danh. We have to thank them and all those whose work has been included. The handbook is recommended reading for anyone looking for environmental friendly and low cost solutions for dealing with water quality issues – it might just provide the answers at a time when the world and its people are facing some very serious environmental problems.

Dick Grimshaw
Founder and Director of The Vetiver Network International.
December 2015

FOREWORD

This work provides a comprehensive view about the great potential of Vetiver System Technology (VST), mainly based on Vetiver grass (*Chrysopogon zizanioides*), in remediating a wide range of polluted water and contaminated land. It is useful for government, academia, industry and individuals in decision-making about the application of VST for the environmental protection. Vetiver grass has been demonstrated through extensively scientific research that it possesses nearly all the characteristics of an ideal plant for the phytoremediation of water and soils contaminated by heavy metals and organic pollutants. It has a dense and massive root system, exhibits rapid growth, has high biomass production and high accumulation rate of heavy metals in roots (particularly lead, zinc and iron) and nutrients (specifically nitrogen and phosphorous), absorb and promote biodegradation of organic wastes (such as 2,4,6-trinitroluene, phenol, benzo[a]pyrene, atrazine, diuron and tetracycline), and also possesses high tolerance and adaptability to a wide range of weather, environmental and soil conditions. Particularly, the successful field studies of applying Vetiver over the world, have further confirmed that Vetiver grass is a right choice for phytoremediation of certain contaminated water and soils. The strong evidences from the scientific research and field studies presented in this handbook have contributed to the promotion of using VST for the environmental protecting purposes. It in turn profits not only users but also local communities where Vetiver is cultivated. It is due to the fact that VST is an effective, inexpensive, easily implemented and environmentally friendly approach, and may generate additional income sources for users and communities through handicraft production, animal feeds and cooking fuels.

Paul Truong[*] and Luu Thai Danh[**]
Authors

---

[*]TVNI Technical Director, Brisbane, Australia
Email: paultruong@vetiver.org
[**]College of Agriculture and Applied Biology, Cantho University, Vietnam
Email: ltdanh@ctu.edu.vn

CONTENTS

I.	INTRODUCTION	1
II.	HOW THE VETIVER SYSTEM TECHNOLOGY WORKS	1
III.	SPECIAL FEATURES OF VETIVER SUITABLE FOR ENVIRONMENTAL PROTECTION PURPOSES.	2
	3.1. Morphological attributes	2
	3.1.1. Vetiver roots	2
	3.1.2. Vetiver shoots	3
	3.2. Physiological characteristics	5
	3.2.1. Tolerance to extreme weather conditions	5
	3.2.2. Tolerance to fire	6
	3.2.3. Tolerance to adverse soil conditions	7
	3.2.4. Tolerance to a wide range of heavy metal pollutants	8
	3.2.5. Tolerance to agrochemicals, organic pollutants and antibiotics	10
	3.2.5.1. Atrazine	10
	3.2.5.2. Antibiotics	11
	3.2.5.3. Phenol	12
	3.2.5.4. 2,4,6-trinitroluen (TNT)	13
	3.2.5.5. Aromatic compounds (Benzo[A]pyrene)	14
	3.2.5.6. Crude oil	14
	3.2.5.7. Dioxin	15
	3.2.6. Tolerance to fly ash	17
	3.2.7. Tolerance to extremely high level of nutrients in water and soil	17
	3.2.8. High removal rate of nutrients in water and soil	18
	3.2.8.1. Nitrogen and phosphorous	18
	3.2.8.2. Aluminium	20

		3.2.8.3.	Boron	20
		3.2.8.4.	Fluor	20
	3.2.9.	High transpiration rate		21
	3.2.10.	Better performance than other plant species		21
3.3.	Agronomic characteristics			22
	3.3.1.	High biomass production		22
	3.3.2.	Minimal competition for nutrient and moisture		23
	3.3.3.	Strong symbiotic association with microorganisms in the rhizosphere		23
	3.3.4.	Highly resistant to disease and pests		23
	3.3.5.	Pest control		24
3.4.	Other important characteristics			25
	3.4.1.	Vetiver is sterile and non-invasive		25
	3.4.2.	Long life-span		26

IV.	COMPUTER MODELS APPLIED FOR WASTE-WATER TREATMENT OF VETIVER GRASS			26
V.	PREVENTION AND TREATMENT OF POLLUTED WATER			28
	5.1.	Treatment of sewage effluent		29
		5.1.1.	Disposal of domestic sewage effluent	25
		5.1.2.	Disposal of community sewage effluent	34
		5.1.3.	Disposal of municipal sewage effluent	38
			5.1.3.1. Small scale application	38
			5.1.3.2. Large scale application	40
			5.1.3.3. Catchment scale application	42
	5.2.	Disposal of industrial wastewater		44
		5.2.1.	Wastewater from a gelatin factory and a beef abattoir	44
		5.2.2.	Wastewater from intensive farms	46
		5.2.3.	Wastewater from a seafood processing factory	47

		5.2.4.	*Wastewater from a small paper factory*	48
		5.2.5.	*Wastewater from a tapioca mill factory*	48
		5.2.6.	***Phenol contaminated water from illegal dumping of industrial waste***	49
		5.2.7.	*Wastewater from oil processing factory*	50
		5.2.8.	*Wastewater from a palm oil mill factory*	50
		5.2.9.	*Wastewater from an aluminium manufacturer*	51
		5.2.10.	*Wastewater from a fertilizer company, quarry industry and a public refuse dumpsite*	52
		5.2.11.	*Wastewater from mixture of laboratory and sewage sources*	53
	5.3	**Disposal of landfill leachate**		55
		5.3.1.	*Landfill leachate disposal in Australia*	55
		5.3.2.	*Landfill leachate disposal in Mexico*	56
		5.3.3.	*Landfill leachate disposal in Morocco*	57
		5.3.4.	*Landfill leachate disposal in the United States*	58
		5.3.5.	*Landfill leachate disposal in Iran*	59
	5.4.	**Municipal landfill leachate seepage control**		60
	5.5.	**Reducing toxic elements in irrigation water**		62
VI.	**PREVENTION, TREATMENT AND REHABILITATION OF MINING WASTES AND CONTAMINATED LAND**			63
	6.1.	**Gold mine**		66
	6.2.	**Coal mine**		74
		6.2.1.	*Overburden*	74
		6.2.2.	*Tailings*	75
	6.3.	**Bentonite mine**		78
	6.4.	**Bauxite mine**		80
	6.5.	**Copper mine**		85

	6.6.	Lead/zinc mines	89
	6.7.	Iron ore mine	93
	6.8.	Ammonia and nitrate contaminated land	95
	6.9.	Hydrocarbon contaminated land	98
	6.10.	Agricultural chemicals contaminated land	98
VII.	REFERENCES	100

I. INTRODUCTION

In the course of researching the application of Vetiver grass's extraordinary attributes to soil and water conservation, the grass was also found to possess unique physiological and morphological characteristics which is particularly well suited to environmental protection, particularly in the prevention and treatment of contaminated water and land. These remarkable characteristics include a high level of tolerance to elevated and even toxic levels of salinity, acidity, alkalinity, sodicity, and a whole range of heavy metals and agrochemicals, as well as exceptional ability to absorb and tolerate elevated levels of nutrients and uptake of large quantities of water in the process of producing a massive growth under wet conditions.

The application of the Vetiver System Technology (VST) as a phytoremediation tool for environmental protection is an innovative approach that has tremendous potential. VST is a natural, green, simple, practicable and cost-effective solution. Most importantly, Vetiver's leaf by-product offers a range of uses from handicrafts, animal feeds, thatches, mulch and fuel, to name just a few.

Its effectiveness, simplicity and low cost make the VST a welcome partner in the many tropical and subtropical countries for domestic, municipal and industrial wastewater treatment; and contaminated land and mining wastes phytoremediation and rehabilitation.

II. HOW THE VETIVER SYSTEM TECHNOLOGY (VST) WORKS

VST prevents and treats contaminated water and land in the following ways.

Preventing and treating contaminated water
- Eliminating or reducing the volume of wastewater,
- Improving the quality of wastewater and polluted water,
- Absorbing nutrients, heavy metals and other pollutants.

Preventing and treating contaminated land
- Controlling offsite pollution,

- Phytoremediation of contaminated land,
- Trapping eroded materials and trash in runoff water,
- Absorbing nutrients, heavy metals and other pollutants.

III. SPECIAL VETIVER FEATURES SUITABLE FOR ENVIRONMENTAL PROTECTION PURPOSES.

Special characteristics of Vetiver are directly applicable to environmental protection purposes, including the following morphological, physiological and agronomic attributes.

3.1. Morphological attributes

3.1.1. Vetiver roots

The success of using Vetiver for phytoremediation of contaminated soils and water depends on the interaction between its roots and contaminated bodies. Vetiver possesses a lacework root system that is abundant, complex, and extensive (Figure 1). The root system can reach 3-4 meters in the first year of planting (Hengchaovanich, 1998) and acquires a total length of 7 meters after 36 months (Lavania, 2003). The features of the root system support its survival under extreme drought conditions as it can utilize deep soil moisture. The root system prevents Vetiver from dislodgement under high velocity flows (Hengchaovanich, 1999; Hengchaovanich and Nilaweera, 1998). However, the grass certainly may not penetrate too far down into the groundwater table. Therefore at locations with high groundwater level, its root system may not be as long as in drier soil (Van and Truong, 2008). Furthermore, mmost Vetiver roots are very fine with an average diameter of 0.66 mm (range from 0.2-1.7 mm) (Cheng et al., 2003). The vertical growth rate of Vetiver root reaches a plateau of approximately 3 cm per day at the soil temperature of 25°C. At the higher soil temperature, the root extension rate is higher but not statistically significant. At the lower soil temperature (13°C), underground root growth was still detected indicating that Vetiver is not dormant at this temperature (Wang, 2000). The horizontal spreading of lateral roots was in the range of 0.15-0.29 m with an average of 0.23 m (Mickovski et al., 2005). Similarly, root growth of Vetiver was measured about 25 cm wide in the study of Nix et al. (2006). After 8 months of cultivation, Vetiver produced 0.48 kg of dry roots per plant. The peculiarity of Vetiver's root system ensures high contact surfaces with soil particles and contaminants resulting in efficient phytoremediation of contaminated soils and wastewater.

Figure 1. Massive, penetrating and deep root systems.

3.1.2. Vetiver shoots

Vetiver has an unusual leaf and shoot architecture. Unlike most of other grasses, Vetiver has a V shape leaf with a prominent mid rib, which can control the opening or closing of the leaf. Under moist or wet conditions, the leaves open up, resulting in higher transpiration rate. Liao et al (2003) found that leaf blades of Vetiver grown in wetland became thinner and the density of stomata increased – an ideal combination for wastewater disposal; but under dry conditions, the leaves close up resulting in lower transpiration rate to conserve moisture (Figure 2), so it is very drought tolerant. The stiff and erect shoots form a dense funnel shape canopy with leaves varying between 45° and 135°, not flat or horizontal like broadleaf plants or most grasses (Figure 3). This shoot architecture has several important implications: the longer sunlight interception of individual leaf as the sun moves from east to west, sunlight interception from both sides of the individual leaf, exposing most of the leaves simultaneously to sunlight, with minimal shading of leaves within the canopy as most other plants. Consequently, larger leaf surfaces of Vetiver are exposed to sunlight over longer time for photosynthesis, leading to better growth as compared to other plant species.

Figure 2. V shape leaves with a prominent mid rib that can open or close leaf blade.

Figure 3. Stiff and erect shoots with 45°-135° angle (left), forming a thick hedge when planting close together (right).

3.2. Physiological characteristics

Extensive research on physiological characteristics of Vetiver over past decades has shown that it is an excellent candidate for a wide range of phytoremediation applications according the special aspects presented below.

3.2.1. Tolerance to extreme weather conditions

Firstly, Vetiver is highly adaptable to extreme weather conditions. It can thrive and survive under the prolonged drought, flood as well as extreme hot and cold weather. The extensive and long root of Vetiver, mentioned above, can utilize deep soil moisture supporting the survival of Vetiver grass up to 6 months under drought condition (Figure 1). Moreover, Vetiver grass is considered as a hydrophyte (wetland plant) due to its well-developed sclerenchyma (air cell) network. Consequently, Vetiver can thrive under hydroponics conditions. Vetiver was demonstrated to be tolerant to the complete submergence for more than 120 days (Xia et al. 2003). Similarly, Vetiver can survive more than 3 months under muddy water in a trial conducted in 2007 to stabilize the Mekong riverbank in Cambodia (Toun Van, pers.com.). Under partial submergence, it can stand up to 8 months in a trial in Venezuela (Figure 4). Furthermore, Vetiver can stand very high temperature up to 55°C in Kuwait (Xia et al., 1999). Additionally, under frosty weather Vetiver top growth was killed but its underground growing points survived (Truong et al., 2008). Vetiver growth was not affected by severe frost at –11°C in Australia and it survived for a short period at –22°C in northern China.

Figure 4. Vetiver survival under prolong drought (left) in Australia (note: all native plants were browned off); and submergence in 25 cm of water for 8 months (right) in Venezuela. Source: www.vetiver.org.

3.2.2. Tolerance to fire

Dry or frosted Vetiver shoots can be burnt readily, but it can survive severe fires and fully recover after burning because its growing point is underground (Figure 5).

Figure 5. Strong recovery of Vetiver growth after the heavy fires in Vanuatu (left) and Australia (right). Source: www.vetiver.org.

3.2.3. Tolerance to adverse soil conditions

Another interesting characteristics of Vetiver is the high tolerance of a wide range of extreme soil conditions, such as high and low pH, high aluminum, high salinity, and high sodicity. Glasshouse and field experiments showed that Vetiver can grow well on the soils with pH ranging from 3.3 – 9.5 (Danh et al., 2009). Particularly, Vetiver showed excellent growth on old gold tailings (pH = 2.7) and bauxite mine tailings (pH = 12) in Northern Queensland, Australia (Danh et al., 2012). Vetiver can grow on the soils with aluminum saturation level (ASL) of 68-86%, however the grass did not survive at ASL of 90% with soil pH of 2 (Truong and Baker, 1997). The study in Vanuatu recently indicated that Vetiver can thrive on highly acidic soils with ASL of 87% (Truong, 1999). Vetiver can grow on saline soils with EC_{se} up to 47.5 dS m^{-1}, its salinity threshold is at EC_{se} = 8 dS m^{-1} and soil EC_{se} values of 20 dS m^{-1} reduce yield by 50%. Vetiver grass was also demonstrated to be able to grow in seawater with salinity ranging from 0 - 19.64 dS m^{-1}, equivalent to 0 - 11 ‰ salt (Cuong et al., 2015). For this reason, Vetiver is classified to a group of highly salt tolerant crop and pasture species grown in Australia (Greenfield, 2002). In addition, the growth of Vetiver grass on the soil with exchangeable sodium percentage (ESP) up to 48% was not adversely affected (Bevan et al., 2000), while the value of ESP higher than 15% considered to be strongly sodic (Northcote and Skene, 1972).

3.2.4. Tolerance to a wide range of heavy metal pollutants

One special attribute of Vetiver discovered recently has made it an excellent plant for heavy metal phytoremediation is its ability to highly tolerate and accumulate a wide range of heavy metals. While most vascular plants are highly sensitive to heavy metal toxicity and most plants were also reported to have very low threshold levels for metals in the soils, Vetiver grass can tolerate not only high concentrations of individual heavy metals in soils but also combinations of several heavy metals (Danh et al., 2012). A series of single heavy metal experiments under glasshouse conditions proved that Vetiver has high tolerance to a wide range of heavy metals in soils due to its high threshold levels of these metals in soils (Table 1). Vetiver could survive and grow well on multi-heavy metal contaminated soils under glasshouse conditions with total Pb, Zn and Cu in the range of 1155 - 3281.6, 118.3 - 1583 and 68 - 1761.8 mg kg^{-1}, respectively. Vetiver was also demonstrated to grow well on iron ore tailings containing high concentrations of multi-heavy metals with total Fe, Zn, Mn and Cu concentrations of 63920, 190, 3220 and 190 mg kg^{-1}, respectively (Roongtanakiat et al., 2008). Under field conditions, Vetiver could grow on mine tailing soils containing total Pb, Zn, Cu and Cd of 2078 - 4164, 2472 - 4377, 35 - 174 and 7 - 32 mg kg^{-1}, respectively. Recently, Vetiver grass has been shown to accumulate high content of these metals in its roots and shoots (Table 2). Majority of heavy metals accumulated in the Vetiver roots and only small portions transported into the shoots make Vetiver grass suitable for phytostabilization of heavy metal contaminated soils (Danh et al., 2012).

Table 1. Threshold levels of heavy metals to Vetiver growth based on single element experiment (Danh et al., 2012).

Heavy metals	Threshold to growth of most vascular plants (mg kg^{-1})		Threshold to Vetiver growth (mg kg^{-1})	Vetiver survival under the highest levels of metals reported in the literature (mg kg^{-1} soil)
	Hydroponic	*Soil level*	*Soil level*	
Arsenic	0.02-7.5	2.0	100-250	959
Boron				180
Cadmium	0.2-9.0	1.5	20-60	60
Copper	0.5-8.0	NA	50-100	2600
Chromium	0.5-10.8	NA	200-600	2290
Lead	NA	NA	>1500	10750
Mercury	NA	NA	>6	17
Nickel	0.5-2.0	7-10	100	100
Selenium	NA	2-14	>74	>74
Zinc	NA	NA	>750	6400
Iron				63920 [1]
Manganese				3220 [1]
Uranium				250 [2]

Note: [1] Roongtanakiat et al., (2008), [2] Hung et al., (2012).

Vetiver was recently investigated for its potential in accumulating uranium (U) from four artificially contaminated soils (Hung et al., 2012). The soils were spiked with aqueous solution of uranyl nitrate at four concentrations of U: 0, 50, 100 and 250 mg kg^{-1} dry soil. It was found that Vetiver grew well on the contaminated soils, even at the highest level of added U, and showed no symptoms of toxicity. In addition, the biomass of the grass grown in soils contaminated with high concentration of U was not significantly different from that of the control grass. Vetiver accumulated more U in the roots than in the shoots, and the uptake is dependent on the soil properties. High soil salinity increased the uptake, but organic matters, ferrous, potassium, and clay content reduced the uranium uptake of Vetiver. The lower nutrient content of soils the higher uranium uptake by the plant. It can be concluded that Vetiver is a potential plant for phytoremediation of soils contaminated with uranium.

Table 2. The highest concentrations of heavy metals accumulated in the roots and shoots of Vetiver reported in the literature (Danh et al., 2012).

Heavy metals	Soil condition		Hydroponic condition	
	Roots (mg kg^{-1})	Shoots (mg kg^{-1})	Roots (mg kg^{-1})	Shoots (mg kg^{-1})
Lead	4940	359	≥ 10,000	≥ 3350
Zinc	2666	642	>10,000	>10,000
Chromium	1750	18		
Copper	953	65	900	700
Arsenic	268	11.2		
Cadmium	396 [1]	~ 44	2232	93
Mercury			1310 [2]	
Iron	871 [3]	1197 [3]		
Manganese	552 [3]	648 [3]		
Uranium	28 [4]	164 [4]		

Note: [1] Zhang et al., (2014), [2] Lomonte et al., (2014), [3] Roongtanakiat et al., (2008), [4] Hung et al., (2012).

3.2.5. Tolerance to agrochemicals, organic pollutants and antibiotics

Vetiver has been recently found to be highly resistant to a range of organic pollutants in growing media, including agrochemicals, antibiotics and other organic wastes (Table 3). Particularly, Vetiver was demonstrated to have ability to remove phenol, tetracycline and 2,4,6-trinitroluen (TNT) from growing media.

3.2.5.1. Atrazine

Vetiver can tolerate up to 20 ppm of atrazine for six weeks, even with a maximum bioavailability created by the use of a hydroponic system (Marcacci et al., 2006). It can be explained by the fact that Vetiver possesses the effective detoxifying processes involving conjugation and dealkylation of atrazine in which conjugation clearly dominates on dealkylation. The conjugated atrazine was mainly detected in leaves, while the dealkylate products were found in both roots and leaves. Furthermore, Vetiver roots were demonstrated to be able to sequester atrazine in the lipid content where Vetiver oil could concentrate atrazine. Vetiver oils in the root increase with aging thus atrazine sequestration in roots may increase with time. Because of the constant growth of the root system, some atrazine in the water could be trans-located to the shoot with the transpiration stream, where detoxification occurs. Under soil condition, the plant growth of Vetiver, measured by leaf chlorophyll activity, was not affected by the application of high atrazine concentration, equivalent to 1 mg/L. The reduction of atrazine in Vetiver treated soils was significantly greater than of the control treatment, owing to atrazine accumulation of Vetiver and microbial degradation of atrazine induced by Vetiver roots in rhizosphere (Winter, 1999). It can be concluded that the combination of these Vetiver properties make it an ideal plant for phytoremediation of atrazine and maybe extended to other agricultural and industrial pollutants, such as dioxin.

Table 3. The tolerance of Vetiver to the highest concentrations of organic pollutants in growing media reported in literature.

Organic pollutants		Soil	Hydroponic	References
Agrochemicals				
	Atrazine		20000 µg L^{-1}	1
	Diuron		2000 µg L^{-1}	2
Antibiotics				
	Tetracycline		15 mg L^{-1}	3
Others				
	Phenol		1000 mg L^{-1}	4
	2,4,6-Trinitroluene	80 mg kg^{-1}		5
			40 mg L^{-1}	6
	Benzo[A]pyrene	100 mg kg^{-1}		7
	Petroleum hydrocarbons	5%		8

Note: 1 Marcacci et al., 2006; 2 Cull et al., 2000; 3 Datta et al., 2013; 4 Singh et al., 2008; 5 Das et al., 2007b; 6: Makris et al., 2007b; 7 Li et al., 2006; 8 Brandt et al., 2006.

3.2.5.2 Antibiotics

Vetiver completely removed tetracycline (TC) from all treatments with three concentrations of TC (5, 10, and 15 mg L^{-1}) within 40 days, whereas no significant reduction in the TC concentrations was found in absence of Vetiver grass (Datta et al., 2013).

3.2.5.3. Phenol

Vetiver plantlets grown under aseptic conditions could remove almost all phenol from media with phenol concentration less than 200 mg L^{-1} in a period of 4 days (Singh et al., 2008). Vetiver removed 89%, 76%, and 70% phenol within 4 days as grown in the media of 200, 500 and 1000 mg phenol L^{-1}, respectively. As plant investigated under aseptic conditions without the confounding effect of microorganisms, this study indicated that Vetiver was solely responsible for phenol remediation. However, the study of Phenrat et al. (2015) suggested that phenol degradation by Vetiver involves two phases (Figure 6). The first phase included phyto-oxidation and phytopolymerization of phenol assisted by root-produced H_2O_2 and peroxidase (POD). The second phase was a combination of the first phase with the enhanced rhizomicrobial degradation. Initially, phenol was rapidly detoxified to phenol radicals, followed by polymerization to non-toxic polyphenols or selective polymerization with natural organic matters, which were then precipitated as particulate polyphenols (PPP) or particulate organic matters (POM). After the first phase, the concentration of phenol significantly decreased, while that of PPP and POM greatly increased, as indicated by the increase of particulate chemical oxygen demand. Synergistically, rhizomicrobes intensively grew on the roots of Vetiver grass and participated in microbial degradation of phenol at the lower concentration, increasing phenol degradation rate by more than 4-folds in comparison to phenol degradation rate in the first phase, and by approximately 32-folds compared with phenol removal rate without Vetiver grass. The combined effects of root-assisted phyto-oxidation and phyto-polymerization, and rhizomicrobial degradation resulted in the complete removal of phenol in wastewater.

The feasibility of phenol degradation by vetiver grass was investigated on a floating platform and a wetland with horizontal flow (WHF) at laboratory scale (Phenrat et al., 2015). Phenol was degraded at a rate constant of 9.7×10^{-3} h^{-1} for the floating platform treatment (100 vetiver plants per 35 L of wastewater), and 10×10^{-3} h^{-1} for the WHF treatment (20 vetiver plants over a length and width of 40 and 20 cm, respectively). The phenol degradation rate of vetiver treatments are about 10 times slower than that of ultrasound (advanced engineering techniques) with a rate of 111×10^{-3} h^{-1}. However, phytoremediation of phenol and other hazardous substances by vetiver grass is much more suitable as considering the practicality of this technique and the widespread nature of contamination.

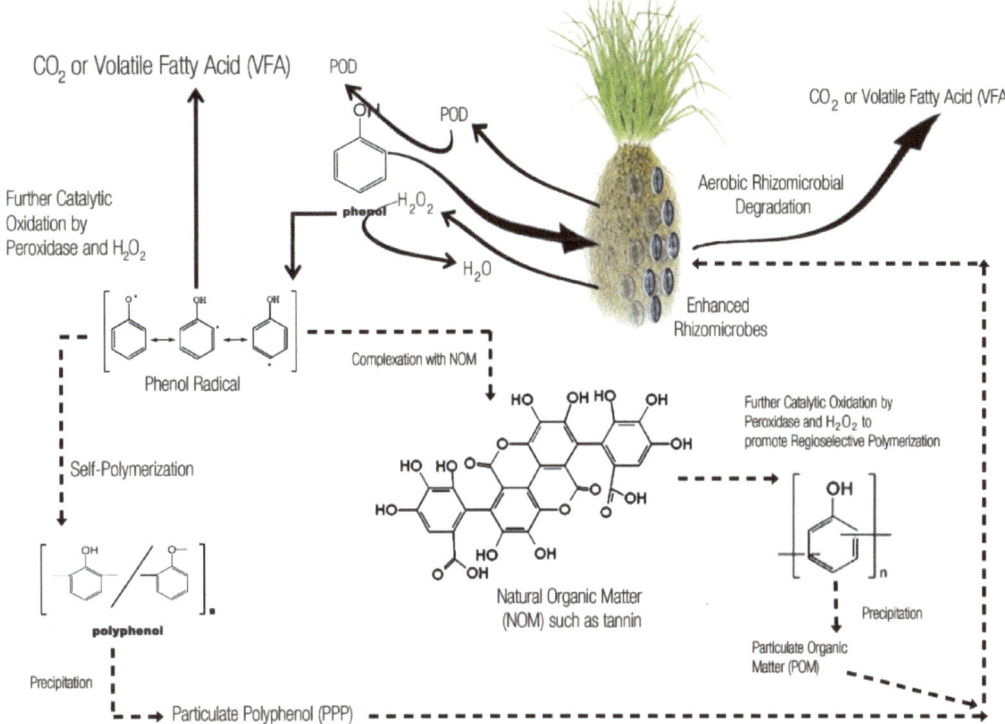

Figure 6: The hypothetical mechanism of phenol degradation by vetiver grass in wastewater (Source: Phenrat et al., 2015).

3.2.5.4. 2,4,6-trinitroluen (TNT)

Under hydroponic condition, Vetiver was demonstrated to have high affinity for TNT by nearly complete removal of TNT from 40 mg TNT L^{-1} solution after 8 days of treatment (Makris et al., 2007a). TNT removal kinetic of Vetiver was significantly increased by the addition of urea as a chaotropic agent (Makris et al., 2007b). No TNT was detected either in roots or shoots, but three major TNT metabolites were found in the roots, but not in the shoot, indicating TNT degraded by Vetiver roots. Similarly, Vetiver could reduce 97% of TNT in soil treated with 40 mg kg^{-1} TNT after 3 days (Das et al., 2010). As the initial TNT concentration was doubled (80 mg kg^{-1}), after 3 and 12 days Vetiver could remove 39% and 88% TNT in soil without urea addition, respectively, however up to 84% and 95% TNT in soil removed with urea amendment, respectively.

3.2.5.5 Aromatic compounds (Benzo[A]pyrene)

In horizontal subsurface flow constructed wetland, Thao Minh Tran et al (2015) showed Vetiver removed 96.8% of phenol and almost 100% of benzene. Under floating raft treatment Vetiver roots removed 91.5% of phenol and almost 96% of benzene

3.2.5.6. Crude oil

In a small trial conducted to remediate contaminated soil at an oil-drilling site in Argentina, results showed that Vetiver can grow on soil heavily polluted by crude oil. Better growth was observed when Vetiver was planted on the contaminated soils covered with 5 cm of organic compost or on the mixtures of 70% contaminated soils and 30% compost, and 50% contaminated soils and 50% compost. After 5 months of cultivation, young and healthy Vetiver roots that grew into the contaminated soil layer gradually changed soil color from red-brown to dark grey (Figure 7) and reduced the petroleum odor from the soil. The results imply a drastic decrease in the concentration of hydrocarbons in the contaminated soil and maybe an increase in soil organic content and microbial activity. From this preliminary study, Vetiver can be considered as an excellent alternative for the treatment of soil contaminated with hydrocarbons (M.T.D. S. Ferrari pers.com).

Figure 7. The colour of the soil layer changed when the roots penetrated into the contaminated soil.

In a study aimed at determining the tolerance of Vetiver to crude oil contaminated soil and its ability to stimulate biodegradation of petroleum hydrocarbon in soil, Brandt et al (2006) in Venezuela found that after 6 months, tiller number of plants grown in soils contaminated with 5% of Venezuelan heavy crude oil war higher than that of control. Despite significant reduction in biomass, plant height and root growth in the presence of crude oil, Vetiver shoots did not show any signs of toxicity. As for the degradation of total oil and grease in soil, no significant decrease in the presence of Vetiver was detected. It can be concluded that the cultivation of Vetiver on oil-polluted sites in Venezuela is considered to be useful. Firstly, Vetiver planting provides soil erosion control thus avoiding offsite pollution. Moreover, if planted on slightly oil-contaminated soil it could ameliorate soils for subsequent cultivation of remedial species.

Tolerance level and rehabilitation potential of Vetiver grass on various processed industrial oils were studied in a small glasshouse trial in Australia. Soil samples were collected from a gold mine in Queensland between 2001- 2007. Six months after planting, it was found that (Truong, pers.com):
- Diesel fuel is highly toxic to Vetiver growth. Vetiver could not survive at 50% mixture of highly contaminated diesel fuel. Diesel is even more deadly if sprayed on leaves,
- Vetiver is moderately tolerant to hydraulic oil,
- Vetiver is highly tolerant to degraded/oxidized hydraulic oil.

3.2.5.7 Dioxin

Anticipating Vetiver exerting a similar effect on dioxin as it did on atrazine, a dioxin phytoremediation project funded by the Vietnamese Ministry of Natural Resources and Environment was initiated at Bien Hoa Air base, Dong Nai Province, Vietnam in 2014 (Huong et al, 2015). Dioxin is a manufacturing contaminant present in Agent Orange that is a 1:1 mixture of 2,4,5-T and 2,4-D herbicide. Dioxins are a family of chemicals that have been implicated to cause serious health effects in humans, including birth defects, rashes, psychological symptoms and cancer. During the Vietnam War, approximately 76 million liters of Agent Orange herbicide were sprayed over 1.82 million ha of Southern Vietnam. A Luoi Valley (65 km west of Hue, near the Laos border) and Bien Hoa air base (near Saigon) were two main stockpile/storage facilities of Agent Orange during the war. The overall objective is to treat the contaminated land at Bien Hoa air base, and more specifically to determine whether Vetiver can grow on such highly contaminated soil and whether it can breakdown dioxin as it did with atrazine.

After preliminary investigation of the dioxin level in this area, Vetiver was cultivated on an area of 300 m^2 with a moderate dioxin-contaminated level (about 1000 – 2000 ppt TEQ). The two main objectives of this project were to investigate:

- The capability of Vetiver grass in phytostabilization of dioxin-contaminated sites, preventing its offsite contamination; and
- Its effectiveness in the bioremediation of the dioxin-contaminated soils.

After four months of cultivation, Vetiver grass grew well on poor quality and moderately toxic chemical/dioxin contaminated soil, with and without soil supplement. Some plants started flowering in week 16, indicating that Vetiver was well established on this kind of contaminated soil. These results demonstrated that the first objective - capability of Vetiver grass in phytostabilization of dioxin-contaminated sites was achieved just four months after planting. This investigation is in progress and the final results are expected in March 2016. If outcomes are promising, large-scale implementation of Vetiver grass technolgy to rehabilitate moderately toxic chemicals/dioxins contaminated soils in Vietnam can be applied (Figure 8).

Figure 8. Vetiver 5 months after planting, with soil supplement (left); and without soil supplement (right).

3.2.6. Tolerance to fly ash

Coal-based power generation is a principal source of electricity in many countries. About 15–30% of the total amount of residues generated during coal combustion is fly ash (FA). Part of FA is re-used in in cements, concrete, bricks, wood substitute products, soil stabilization, road base/embankments and consolidation of ground, land reclamation, and as a soil amendment in agriculture (Asokan et al., 2005; Jala and Goyal, 2006). The rest is disposed in landfills on open landfills that are under pressure from environmental concerns and increasingly stringent environmental regulations. Vegetative reclamation of ash ponds is an economical and effective solution to reduce their environmental impacts, such as preventing fugitive dust emission, controlling soil erosion, stabilizing the surface areas of ash, preventing the potential ground water contamination and finally adding vegetation cover that is vital in the long run. However, FA often inhibits plant growth owing to its characteristics, such as alkalinity, nutrient deficiencies, toxic heavy metal contents, and poor physical structure. Therefore, selection or screening of plant species which are tolerant to toxic levels of heavy metals has attracted much attention in the treatment of the abandoned fly ash dump (Das and Adholeya, 2009).

Vetiver was selected and investigated for the capacity of remediating fly ash-soil amendments (0, 25, 50 and 100%) over a period of 18 months in a pot experiment (Ghosh et al., 2015). The amendments and their respective leachates were subjected to metal analysis to understand the role of heavy metal induced genotoxicity. The roots and leaves of Vetiver grown in different amendments were subjected to metal estimation to help understand the extent of remediation of fly ash under the influence of Vetiver. The study revealed the marked decrease in concentration of heavy metals and the significant decrease in genotoxic potential of the fly ash soil amendments marked by reduction in micronuclei formation, bi-nucleate cells and chromosomal aberrations of Vetiver roots over the period of 18 months. Therefore, Vetiver can be used as an excellent candidate for remediation and restoration of fly ash dumpsites.

3.2.7. Tolerance to extremely high level of nutrients in water and soil

Vetiver has been demonstrated to highly tolerate and accumulate high concentrations of nitrogen (N) and phosphorous (P) - **the main elements causing water pollution** (Figure 9). The application of up to 10,000 and 1,000 kg ha^{-1} year^{-1} of N and P, respectively, did not adversely affect Vetiver growth, however insignificant Vetiver growth response was observed at rates higher than 6,000 and 250 kg ha^{-1} year^{-1} of N and P, respectively (Wagner et al., 2003).

Figure 9. High N and P removal capacity of Vetiver: blue green algae infested waste water (left) with high nitrate (100 mg L^{-1}) and phosphate (10 mg L^{-1}), same effluent after 4 days of treatment with Vetiver (right) reducing N and P level to 6 and 1 mg L^{-1}, respectively. Algal infestation was eliminated from the effluent.

3.2.8. High removal rate of nutrients in water and soil

3.2.8.1. Nitrogen and phosphorous

Vetiver is superior in terms of N and P removal as compared to other grasses (Figure 10). Under hydroponic condition with a sewage effluent flow rate of 20 L min^{-1} through Vetiver roots, one square meter of Vetiver can treat 30,000 mg of N and 3,575 mg of P in eight days (Hart et al., 2003). In this application, Vetiver out-performed other crops and pasture plants, such as Rhodes grass, Kikuyu grass, green panic, forage sorghum, rye grass and eucalyptus trees (Truong, 2003). Vetiver reduced total N and P of the polluted river water (initial concentrations of 9.1 and 0.3 mg L^{-1}, respectively) by 71 and 98%, respectively after 4 weeks of treatment (Zheng et al, 1997). Vetiver could remove up to 740 kg N ha^{-1} and 110 kg P ha^{-1} over 3 months at a nutrient-rich site and 1,020 kg N ha^{-1} and 85 kg P ha^{-1} over 10 months at a lower nutrient site (Vieritz et al., 2003). In a pot experiment (Smeal et al., 2003), Vetiver was demonstrated to have a very high recovery rate for nitrogen in shoots, but quite low for phosphorous (Table 4).

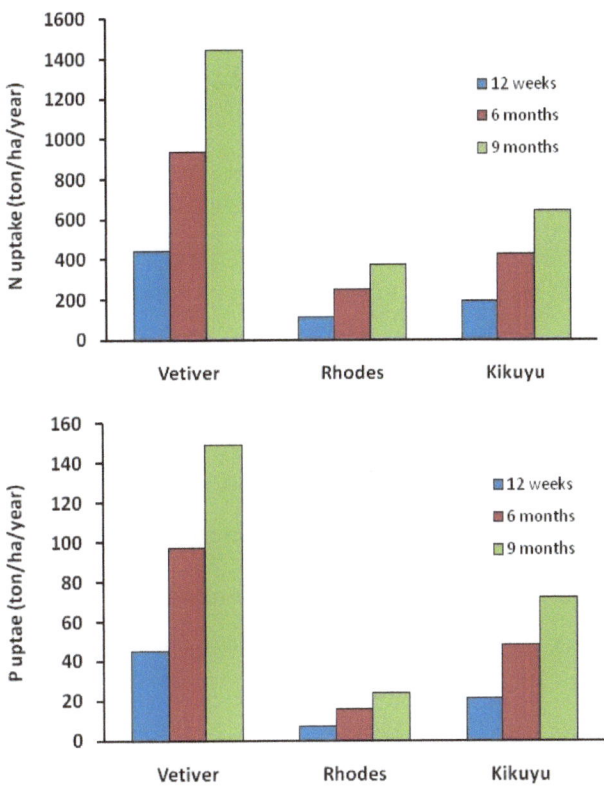

Figure 10. N (above) and P (below) removal of three grasses over time.

Table 4. Recovery rate of N and P by Vetiver.

Treatment	Recovery rate by Vetiver (%)		Recovery in soil (%)	Total
	Shoot	Root		
N (ton ha^{-1} year^{-1})				
2	76.3	20.4	0.3	97
4	72.1	23.1	0.1	95.3
6	67.3	21.2	0.4	88.9
8	56.1	30.0	0.4	86.5
10	46.7	17.0	0.1	63.8
P (kg ha^{-1} year^{-1})				
250	30.5	23.3	46.3	100
500	20.5	14.6	48.7	83.8
1000	16.5	14.2	40.8	71.5

3.2.8.2. Aluminium

Vetiver was demonstrated to be a good candidate for the treatment of aluminium (Al) contaminated wastewater. In a study to quantitatively verify the adsorption of Al from the contaminated industrial wastewater and to evaluate the potential of adsorption of three macrophyte species (Vetiver, *Scirpus lacustris*, *Typha latifolia*), results showed that at Al concentration of 20% *Scirpus lacustris* and Vetiver had the best removal efficiencies (99 and 98%, respectively). Vetiver had very good performance reaching 94% of removal efficiencies at Al concentration of 70%. Maximum adsorption of Al during the phytoremediation process for all species occurred in the first 24 hours. Based on the statistical analysis, the initial water pH was found to be an important factor in the performance of the process (Aldana et al., 2013).

3.2.8.3. Boron

The pioneering study on the ability of Vetiver to remove boron was carried out by Angin et al. (2008) Vetiver was grown in a series of pots that were artificially contaminated with B (0 - 180 mg B kg^{-1}). Boron addition did not impact dry matter yield. After 90 days of experiments, plants were harvested for chemical analysis. The concentration of B accumulated in roots and shoots increased with the level of B in soils. The level of B in Vetiver roots was greater than in shoots: the treatment of 180 mg B kg^{-1} resulted in about 28 mg B kg^{-1} DW in roots while shoots contained about 17 mg B kg^{-1}.

3.2.8.4. Fluor

Vetiver was tested for its ability to remove fluoride from the contaminated water at a community in Guarataro, Yaracuy, Venezuela (Ruiz et al., 2013). This community has serious public health problems, due to the consumption of groundwater contaminated with high level of fluor, which exceeding the limits set by the government. It results in dental fluorosis, characterized by alterations of enamel and in some instances gingival and alveolar injury. 93% of the population of this community suffer dental fluorosis, especially school-children. At the beginning of the experiment, Vetiver had a positive effect in reducing fluoride in water (from 2.72 to 2.22 mg L^{-1}). However, Vetiver did not show a significant effect in decreasing the concentration of fluoride in water in subsequent analysis. Furthermore, the chemical analysis of the plant tissue showed significant absorption of fluoride, demonstrating that Vetiver plant can accumulate this element in its tissues, but the accumulation is not significant for the treatment of fluoride contaminated water. As for nitrogen and phosphorous, Vetiver reduced these pollutants over 90%, demonstrating high efficiency of Vetiver system in removing nutrients.

3.2.9. *High transpiration rate*

A further peculiar aspect of Vetiver is its high transpiration rate, which plays a key role in the phytoremediation of wastewater. It is due to the fact that the plant must transpire enough water from the growing media in order to take up the contaminants effectively (Vose et al., 2004). Truong and Smeal (2003) established a correlation between water use (soil moisture at field capacity) and dry weight (DW) yield of Vetiver. For 1 kg of dry shoot biomass, Vetiver would use 6.86 L day^{-1} of water. The 12 week old Vetiver with dry matter yield estimated at 40.7 t ha^{-1} at the peak of its growth cycle would potentially use 279 KL ha^{-1} day^{-1}. As comparing with other wetland plants, such as *Iris pseudacorus, Typha spp., Schoenoplectus validus, Phragmites australis*, Vetiver has the highest water use rate (Cull et al., 2000). For example, at the average consumption rate of 600 ml day^{-1} pot^{-1} over a period of 60 days, Vetiver used 7.5 times more water than *Typha*.

3.2.10. *Better performance than other plant species*

Literature search on the use of different macrophytes, (published between 1997 to 2014) that have been used for industrial and domestic wastewater treatment, such as pig farm, dairy, sugar factory, textile, tannery, septic tank, domestic, municipal, black water, grey water, river water and lake water, showed that Vetiver is either equally and often more effective in treating wastewater than other macrophytes such as *Cyperus alternifolius, Cyperus exaltatus, Cyperus papyrus, Phragmites karka, Phragmites australis, Phragmites mauritianus, Typha latifolia, Typha angustifolia, Eichhornia crassipes* (Water Hyacinth)*, Iris pseudacorus, Lepironia articutala* and *Schoenoplectus validus*. For example, Le Viet Dung (2015) demonstrated the superiority of Vetiver over Water Hyacinth in treating Biological Oxygen Demand (BOD), pH and nutrient content. While Vetiver continued to grow vigorously, Water Hyacinth died after 8 days in the pig farm wastewater, which had relatively high BOD. Results shows that Vetiver significantly reduced BOD level was by 40% (from 245.80mg/L to 146,37mg/L) after 32 days treatment, while the reduction in Water Hyacinth and control were 21% and 19% respectively. Xia (1997) also found that water hyacinth all died at BOD of 120.8mg/L.

3.3. Agronomic characteristics

3.3.1. High biomass production

Vetiver has a fast growing rate and high biomass production that are two important factors determining its great potential for phytoremediation. Vetiver is a C_4 plant that has high rate of photosynthesis at high light intensities and high temperatures due to the increased efficiency of photosynthetic carbon reduction cycle (Hatch, 1987). Consequently, it has high growth rate, as indicated by the high radiation use efficiency (RUE) of 18 kg ha^{-1} per MJ m^{-1} (Vieritz et al., 2003). The RUE of Vetiver is comparable with other C_4 high biomass producing grasses such as maize (*Zea mays* L.) and sugarcane (*Saccharum officinarum*) which present 16 and 18 kg ha^{-1} per MJ m^{-1}, respectively (Muchow et al., 1990; Inman-Bamber, 1974) and much higher than the RUE of C_3 grasses such as coastal couch grass (*Cynodon dactylon*): 5.3 kg ha^{-1} per MJ m^{-1} (Burton and Hanna, 1985). High growth rate results in high biomass production of Vetiver, about 100 tons of dry matter ha^{-1} year^{-1} under tropical hot and wet conditions (Truong, 2003). Vetiver biomass production was demonstrated to be higher than that of tropical and subtropical pasture crop (Figure 11). Under sub-tropical weather, Vetiver grass still produces relatively high biomass of 10- 20 tons after 5-6 months of cultivation (Shu et al., 2004; Yang et al., 2003), owing to the fact that Vetiver retains high activity of the key enzymes involved in photosynthesis (NADP-MDH and NADP-MET) under temperate climates (Bertea and Camusso, 2002).

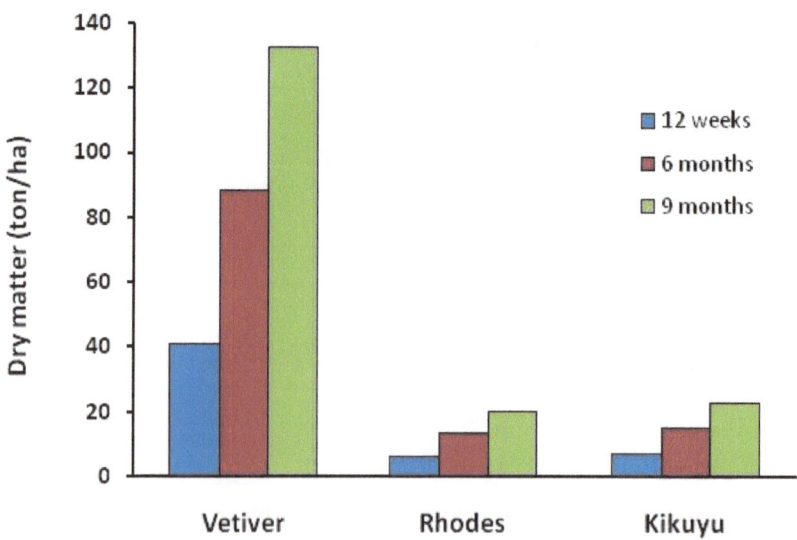

Figure 11. Potential dry matter yield of three grasses over time.

3.3.2. Minimal competition for nutrient and moisture

As shown in Figure 1, most of Vetiver roots grow vertically, especially in the first 30 - 40 cm, hence providing minimal competition for nutrient and soil moisture to accompanying crops. Even at depth, the horizontal spreading of lateral roots was in the range of 0.15 - 0.29 m with an average of 0.23 m (Mickovski et al., 2005). Similarly, Vetiver roots spread about 0.25 m wide in the study of Nix et al. (2006). Hence, Vetiver has been widely used for windbreak to vegetable crops in China, orchard trees in Australia, companion crop to mung bean crop in India and as close as 0.15 m in vegetable garden in Senegal (C.Juliard, pers.com).

3.3.3. Strong symbiotic association with microorganisms in the rhizosphere

Vetiver can survive and grow on soils with very low fertility, particularly organic matter, nitrogen and phosphorous. It is due to the fact that Vetiver can establish a strong symbiotic association with a wide range of soil microorganisms in the rhizosphere (Siripin, 2000; Monteiro et al., 2009; Leaungvutiviroj et al., 2010). These microbes provide nitrogen (nitrogen fixing bacteria), phosphorous (phosphate solubilizing bacteria and fungi, mycorrhizal and cellulolytic fungi) and plant growth hormones (plant growth regulator bacteria) for Vetiver development. Up to 40% of nitrogen content in Vetiver was obtained from the symbiotic association between Vetiver root system and 35 different nitrogen fixing bacterial strains (Siripin, 2000). A large number of bacterial strains were found in the Vetiver rhizosphere being responsible for nitrogen fixation (48 strains), IAA production (46 strains) and phosphate-solubilization (49 strains) in which 25 bacterial strains were determined to involve in three plant growth promoting characteristics (Monteirio et al., 2009). Furthermore, Vetiver also improves the soil quality, in term of nutritional, physical and biological properties, through its symbiotic association with soil micro-organisms (Materechera, 2010; Leaungvutiviroj et al., 2010). Particularly, under the pressure of hostile conditions of heavy metal contaminated soils, the cultivation of Vetiver can increase the microbial populations in such soils.

3.3.4. Highly resistant to disease and pests

Up to date, there are no major reports related to the diseases and pests of Vetiver grass on the world (Danh et al., 2012), except Vetiver is susceptible to the leaf blight caused by *Curvularia trifolii* (http://plants.usda.gov/plantguide/pdf/pg_chzi.pdf). However, there are occasional reports on infestation by *Fusarium* fungus in Colombia and West Papua (Indonesia), Cicadas in New Zealand, stem borer (*Chilo spp*) on planting near rice field in Vietnam, army worms on planting near sugar cane field in Australia, and Hemiptera sucking bug in Venezuela.

3.3.5. *Pest control*

Vetiver planting for environmental protection purposes also acts as a pest control measure to adjacent cropping lands - an extra benefit to the environment. An Integrated Pest Management technique known, as "the push-pull system" using Vetiver grass for crop protection against insect pest was first developed in South Africa (Berg, 2006). In both laboratory and greenhouse studies, the stem borer moths, *Chilo partellus,* were demonstrated to preferably lay their eggs on Vetiver leaves and not on maize. Follow-up studies under greenhouse and field conditions showed that *C. partellus* larval survival on Vetiver was extremely low. With this system, Vetiver grass is used as trap crop to attract the stem borer when planted as an intercrop between the maize. This result can also be applied in sugarcane and rice crops (Figure 12). The number of beneficial arthropod species found on Vetiver was much larger than that on maize during winter and summer. Vetiver has also been used successfully to protect crop and orchard plant against nematode infestation in Australia, Senegal and Thailand.

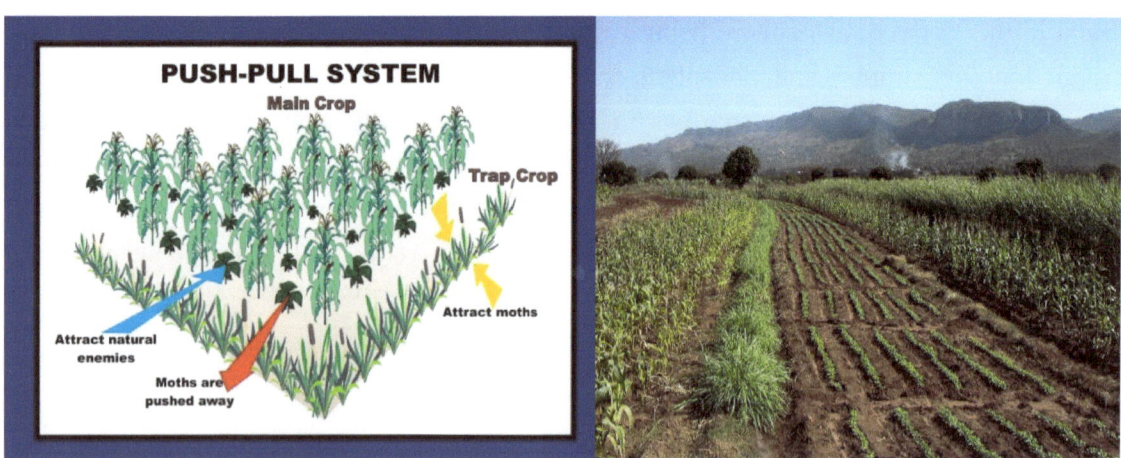

Figure 12. Push-Pull System of pest control and Vetiver planting to protect maize crop in Africa.

In Brazil, a winery used Vetiver hay as mulch for soil and water conservation on its very steep slope vineyard in Minas Gerais state. It was observed that grape vines and fruits were healthier and pest control spray was significantly reduced as the vines and fruits were protected from insect pest (Figure 13).

Figure 13. Vetiver mulch in vineyard protecting grapes from insect pest (A.Pereira pers.com.)

3.4. Other important characteristics

3.4.1. Vetiver is sterile and non-invasive

The chance of Vetiver becoming a weed is very low owing to the fact that the Vetiver grass cultivar deliberately selected for the environmental protection is from south Indian accessions. It produces flowers but sets no seeds and has neither stolons nor rhizomes (Danh et al., 2012). There are several instances illustrating the low invasiveness potential of Vetiver on the world. Vetiver was introduced into Fiji from India for thatching more than 100 years ago, and subsequently it has been widely used for soil and water conservation purposes in the sugar industry for over 50 years. However, there is no sign of invasiveness observed (Truong and Creighton, 1994). Additionally, a study conducted in Australia for 8 years indicated that Vetiver is sterile under various growing conditions (Truong, 2002).

Generally, the cultivated south India accessions have large and strong root systems. These accessions tend towards polyploidy and show high levels of sterility and are not considered invasive. The north Indian accessions, common to the Gangetic and Indus basins, are wild and have weaker root systems. These accessions are diploids and are known to be weedy, though not necessarily invasive. These north Indian accessions are NOT recommended by The Vetiver Network International. It should also be noted that most of the research into different Vetiver applications and field experience have involved the south Indian cultivars that are closely related (same genotype) as Monto and Sunshine. DNA studies confirm that about 60% of Vetiver used for bio-engineering and phytoremediation in tropical and subtropical countries are of the Monto/Sunshine genotype (Adam and Dafforn, 1997).

Recently, under the strict assessment for weed potential - Pacific Island Ecosystems at Risk (http://www.Vetiver.org/USA_PIER.htm), Vetiver was listed as Low Risk with a score of -8. Consequently, the Plant Guide for US Pacific islands published by the US Department of Agriculture – National Resources Conservation Service recommended Monto and Sunshine types to be used for soil and water conservation purpose. (http://plants.usda.gov/plantguide/pdf/pg_chzi.pdf).

3.4.2. Long life-span

Finally, another special characteristic of Vetiver suitable for long term treatment of polluted soils and water is its long life-span, so it is commonly used for demarcation of farm boundary in India and up to 150 years in one case in Vanuatu (Don Miller, pers.com). Therefore after establishment it will grow and develop under adequate maintenance for a long period until phytoremediation completed without further replanting.

IV. COMPUTER MODELS APPLIED FOR WASTE-WATER TREATMENT OF VETIVER GRASS

Vetiver is highly suitable for the treatment of domestic, municipal and industrial wastewater due to its extraordinary attributes such as very high level of tolerance and absorption of pollutants in wastewater, and very high water use rate under wetland conditions (Danh et al, 2009). But most important of all is its capacity to produce a very high biomass under a wide range of climatic conditions and adverse soil conditions. It is due to the fact that the ability of Vetiver grass to remove pollutants and water from the growing medium depends solely on its biomass production, hence the faster and higher biomass production the faster and more effective the treatment process is. Therefore, if the biomass production can be estimated for a certain environment, the efficiency of the treatment process can be predicted and subsequently the land area needed can be worked out reasonably accurately (Truong et al. 2008)

The use of computer models is now an accepted procedure for any effluent management program. During the preliminary design of the dry land irrigation system, a crucial consideration is the amount of land required to dispose of a specified amount of wastewater. The most efficient method to determine the land area required is to simulate a site-specific model for the wastewater disposal.

In Queensland, Australia "Model for Effluent Disposal by Land Irrigation" (MEDLI) is widely used and adopted by the Queensland EPA as a general model for industrial and municipal wastewater management (Truong et al. 2003; Vieritz et al, 2003). But to date, the application of MEDLI in tropical and subtropical Australia has been restricted to a number of tropical and subtropical crops and pasture species, that have much lower treatment capacities than Vetiver grass. Only recently, Vetiver grass applications have been designed using MEDLI, and have resulted in large saving in establishment and operating costs. For example, from the results of MEDLI simulation, the minimum land area required for disposing all wastewater (with the total volume of 475 ML year^{-1}, N and P concentration of 300 and 1 mg L^{-1}, respectively) is 72.5, 104 and 153 ha as growing Vetiver, kikuyu *(Pennesitum clandestinum)* and Rhodes grass (*Chloris guyana*), respectively, at a food processing factory in Queensland (Truong and Smeal, 2003).

MEDLI application is limited to large-scale industrial and municipal wastewater management. In addition, it is based on pasture and crop plant species and it is not suitable for simple or smaller scale using Vetiver grass. Therefore, there was a need for a model that could be applied to sites where MEDLI is not suitable. As a result, Veticon Consulting developed Effluent Disposal by Vetiver Irrigation (EDVI) for situations where MEDLI is not applicable. EDVI is based on some components of MEDLI and the update "Australia Water Balance Model" which is similar to those used in Europe and North America. Particularly, EDVI was designed exclusively for Vetiver grass, using data from extensive R&D results on the effectiveness and capacity of Vetiver grass in treating effluent and leachate in the last 20 years. Major inputs parameters of EVDI needed for the modeling process include long term (50-100 years) and accurate climate data, soil type and depth, ground water level and accurate quantity and quality of the wastewater input and local EPA limits for discharged water (Truong and Truong, 2013).

While the application of VS for large scale projects continues to spread around the world, there is an increasing need for its use in small scale projects to treat low volume domestic and small community wastewater in developing as well as developed countries. To date, all small-scale wastewater treatment projects using Vetiver grass are based on trial and error method and experience. To overcome this, a scientifically based model is needed to convince authorities of its effectiveness and accuracy. Obviously for small-scale application, which produces low volume effluent, the long term and accurate parameters are not easily available or non-existent, hence an accurate determination of the land area needed is very difficult to make. To fulfill this need, Effluent Disposal by Vetiver Irrigation for small volume (EDVI-2) was developed specifically to treat small volume input from individual household and small community sewage effluent and small volume landfill leachate around the world as well as industrial wastewater from small factories or cottage industries, such as coffee farmer or small coffee co-operatives, in Latin America and globally (Truong and Truong, 2013).

V. PREVENTION AND TREATMENT OF POLLUTED WATER

In general, Vetiver System Technology (VST) treats wastewater in the following methods:

1. Land irrigation by:
   - surface flow irrigation or
   - overhead irrigation sprinklers

2. Constructed wetlands:
   - on ground ephemeral wetland or
   - above ground constructed wetland

3. Hydroponics: pontoons or floating platform

4. Natural wetlands

From a technical standpoint, this technology blends numerous disciplines including engineering, hydraulic, hydrology, microbiology, plant physiology/morphology, soil science, agronomy, chemistry and computer science for programming and modeling. The followings are case studies of global applications of Vetiver in treating polluted wastewater from domestic and industrial discharges.

5.1. Treatment of sewage effluent

5.1.1. Disposal of domestic sewage effluent

The first application of the VPT for effluent disposal was conducted in Australia in 1996 for the treatment of the effluent discharge from a toilet block in a park. With the cultivation of about 100 Vetiver plants in an area less than 50 m², the wastewater was completely dried up (Figure 14). Whilst other plants such as fast growing tropical grasses and trees, and crops such as sugar cane and banana failed (Truong and Hart, 2001).

Figure 14. After 6 months of cultivation, 100 Vetiver plants absorbed all discharges from the toilet block.

Groundwater monitoring of the Vetiver cultivation (collected at 2 m depth) showed that after passing through 5 rows of Vetiver, the levels of total N reduced by 99% (from 93 to 0.7 mg L¹), total P by 85% (from 1.3 to 0.2 mg L^{-1}), and fecal coliforms by 95% (from 500 to 23 organisms in 100 mL). These levels are well below the thresholds used by Australian Environmental Authority of total N < 10 mg L^{-1}; total P < 1 mg L^{-1} and *E. coli* < 100 organisms in 100 mL (Figure 15).

In Thailand, comparative studies of Vetiver grown in domestic wastewater from the Royal Irrigation Department community revealed that different ecotypes exhibited different growth and adaptability. Surat Thani ecotype was found to exhibit the highest ability (in percentage) to reduce: nitrate (49.33), bicarbonate (42.66), electrical conductivity (5.81), and total soluble solids (82.78), while Monto cultivar exhibited the highest ability to reduce: biological oxygen demand (75.28), total N (92.48), K (14.00), and Na (3.14). The efficiency of wastewater treatment was found to increase with the age of Vetiver plant, and the highest was reached at 3 months of age (Chomchalow, 2006).

Effectiveness of Vetiver in Reducing N in domestic sewage

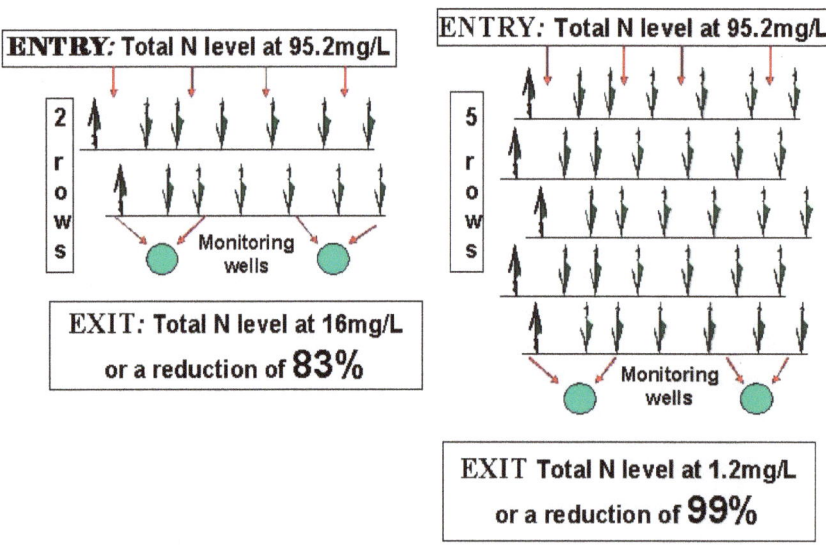

Effectiveness of Vetiver in reducing P in domestic sewage

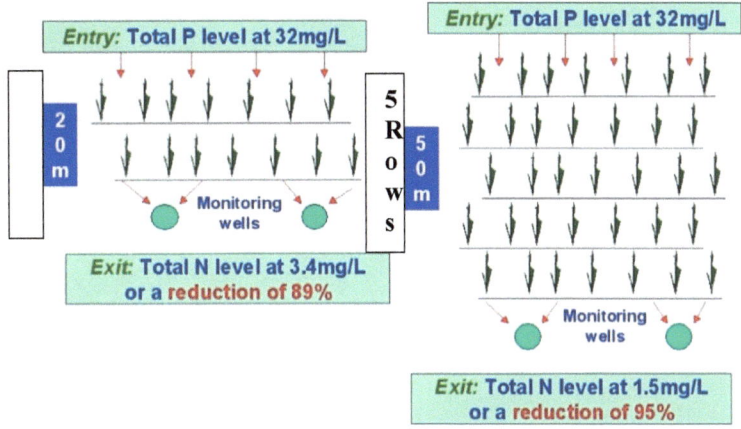

Figure 15. Highly effective in removing Nitrogen and Phosphorus

Recently, the use of Vetiver grass for domestic sewage treatment has been clearly demonstrated in Aceh province, Indonesia, where American Red Cross and Danish Red Cross have built over 3000 houses to resettle the victims of the 2001 Tsunami. Each of these houses has a Vetiver based sewage effluent disposal system (Figure 16). Similar disposal units have been used in Australia (Figure 17), India, Indonesia, Morocco and Papua New Guinea.

Figure 16. Vetiver System for domestic sewage treatment in Indonesia.

Figure 17. Vetiver System for domestic sewage treatment in Australia.

The Vetiver latrine has been recently developed by simply planting Vetiver around a small concrete slab above a pit (Lee, 2013). Instead of bricks and mortar, the long roots of Vetiver stabilize the pit and even remove the environmental contaminates. Above ground, the blades of the grass provide a tall, thick privacy screen that is effectively storm proof. The design is simple enough for households to construct themselves with some basic training. Once the latrine is filled, the slab and seedlings can be transferred over to the next pit location.

The latrine is affordable to the most disadvantaged families. A Vetiver latrine is approximately one twentieth of the price of a traditional latrine because there is no need to transport a large quantity of bricks and construction materials to remote locations for the pit lining and housing and no need for skilled labor for construction (Figure 18).

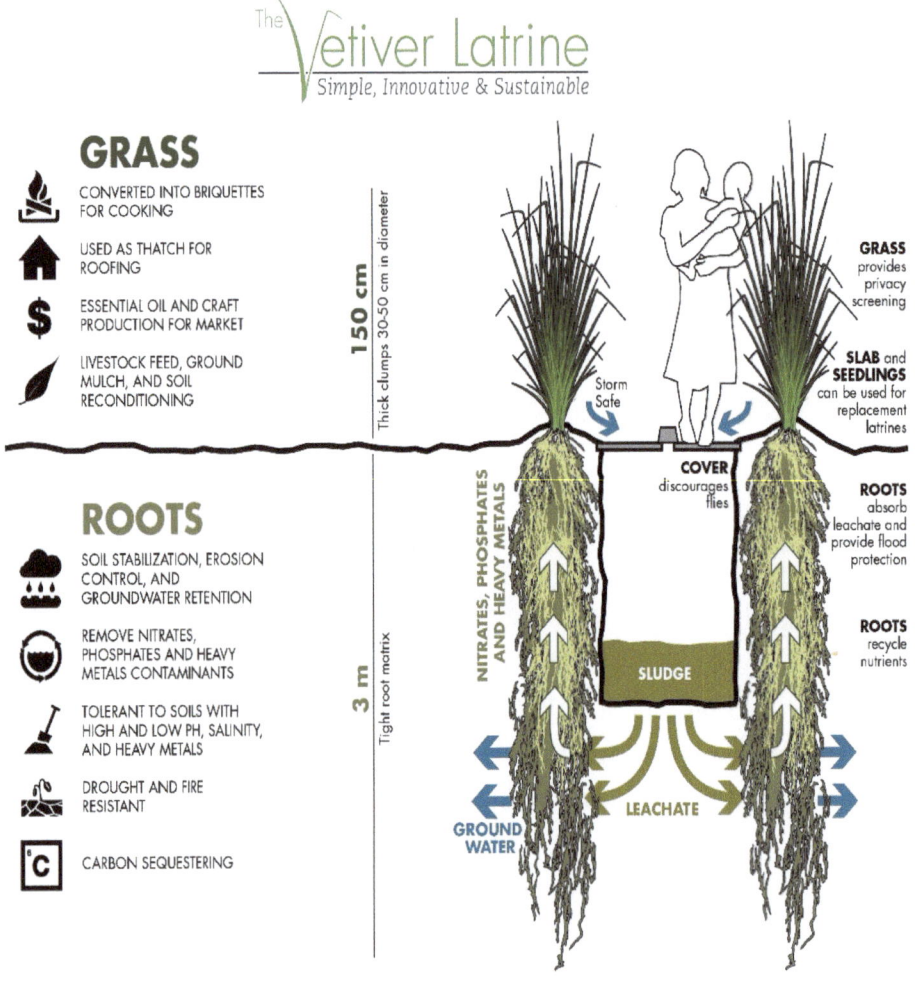

Figure 18. The Vetiver latrine design. Source: Lee, 2013.

5.1.2. *Disposal of community sewage effluent*

Watt Bridge airfield is a small recreational airfield in Queensland, Australia. Vetiver was planted on an area of 100 m² (400 plants grown into 8 rows with 10 m long each, inter-row space of 1 m and 5 plants m⁻¹) to dispose a small volume sewage effluent (Figure 19), with the results presented in Table 5.

Figure 19. Excellent growth after twelve months of cultivation (top), growth exceeding two meters (bottom left), cutting down to 50 cm every three months (bottom right).

Table 5. Water quality before and after treatment with Vetiver at Watt Bridge.

Variables	Inflow	Outflow	Reduction (%)
Average daily flow (L)	1670	Almost nil	Nearly 100
Average total N (mg/L)	68	0.13	Nearly 100
Average total P (mg/L)	10.6	0.152	Nearly 100
Average faecal coliform	>8000	10	Nearly 100

About 400 Vetiver plants were cultivated into eight rows for the treatment of sewage effluent from a storage facility at Refilwein, South Africa. The young plants were watered every three days for the initial three weeks. For the next two months, it received a weekly

irrigation with effluent piped in from the storage pond. After one year of establishment, there was a dramatic difference in growth (Figure 20).

One year after planting, pits were dug to check on water level and *E. coli* concentrations. The pits were dry. From this, it can be deduced that Vetiver has completely disposed of all effluent. The lack of any residual moisture could only mean a probable absence of any pathogenic material had survive their passage through the bed of Vetiver roots (Roley Noffke, pers.com).

Figure 20. Vetiver planting at Refilwe (left) and one year later (right).

A small constructed wetland was installed with the objective of treating 30 KL day^{-1} domestic sewage effluent generated from a community of 300 inhabitants in Tangier on the Mediterranean coast, Morocco (Etienne Richards, pers.com). The effluent was discharged from a septic tank with a capacity of 20 KL. Due to the shortage of land space in urban area, the system consists of a concrete container with 100 m^2 surface area and 1 m in depth, with drain outlet at the bottom to collect samples for analyses (Figure 21). Vetiver was planted at a density of 5 plants m^{-2} (500 plants in total). One month after planting, although not fully mature, the results below show the effectiveness of Vetiver in removing nutrients from the sewage effluent (Table 6).

It is worth noting that the legal maximum levels of recycle water in Morocco for irrigation are: BOD < 120 mg/l; COD < 250 mg/l. So this recycle water can be used for gardening in the town.

Figure 21. Vetiver establishment after one month of cultivation in the concrete container.

Table 6. The water quality before and after Vetiver treatment.

Variables	Inflow	Outflow	Reduction (%)
COD (mg/L)	562	14	97.5
BOD$_{5DAY}$ (mg/L)		<10	
Total N (mg/L)	91	2	97.8
Total P (mg/L)	8.41	0.141	98.3

The Palmavira Resort near Marrakech, Morocco has an existing dysfunctional constructed wetland using Typha. To cope with the increased output of the new extension, the Resort has upgraded its wastewater treatment plant using Vetiver grass instead of Typha (Figure 22).

Figure 22. The dysfunctional wetland using Typha (left) and seedbed ready for Vetiver planting (right).

5.1.3. *Disposal of municipal sewage effluent*

5.1.3.1. *Small-scale application*

The sewage treatment plant for Toogoolawah, a small town in subtropical Australia, was built in the 1970s. The plant was constructed as a primary sedimentation (Imhoff Tank) followed by three sewage storage ponds. The effluent from the ponds was designed to flow down into a swamp before it overflowed into the local creek. The plant construction was based on a very simple design but it was effective. With the recent changes to license conditions imposed by the Environmental Protection Agency (EPA), the plant no longer complies with the license and so an upgrade of the plant was required. Various options were considered such as a nutrient removal plant, a sand filter or a rock filter. These options are expensive and would require expensive ongoing operational costs. The council then considered a Vetiver System treatment system that would take up most of the water, as well as remove nutrients, organic compounds and heavy metals from the sewage effluent (Ash and Truong, 2003).

The Vetiver treatment has composed of two components (Figure 23):
- Hydroponic treatment of effluents in the storage ponds
- Ephemeral wetland

The results of the treatment over the period 2002 - 2004 are summarized in Table 7.

Table 7. Effluent quality characteristics before and after the Vetiver treatment.

Tests (license requirements)	Effluent input	Effluent output
pH (6.5 - 8.5)	7.3 - 8.0	7.6 - 9.2
Dissolved oxygen (2.0 mg L^{-1} minimum)	0 - 2	8.1 - 9.2
5 Day BOD (20 - 40 mg L^{-1} maximum)	130 - 300	7 - 11
Suspended solids (30 - 60 mg L^{-1} maximum)	200 - 500	11 – 16
Total nitrogen (6.0 mg L^{-1} maximum)	30 - 80	4.1 - 5.7
Total phosphorous (3.0 mg L^{-1} maximum)	10 - 20	1.4 - 3.3

Figure 23. Vetiver treatment: hydroponic treatment (top) and ephemeral wetland with area of 1.5 ha (bottom).

5.1.3.2. Large scale application

The Boonah town, near Brisbane, needed to upgrade its sewage treatment plant to comply with new environmental protection law. An Option Planning Report investigated various upgrade options to the plant. The report investigated a number of possible upgrade options for the treatment plant using multi-criteria analysis, determining that a dedicated irrigation disposal system for the effluent was the most suitable, taking into consideration timing and cost factors. The recommended solution was to provide a dedicated pasture grass irrigation solution, which would have required an irrigation area of between 50 and 60 ha.

Subsequently, the preliminary design investigations have been undertaken to develop the dedicated irrigation solution, and in particular a variation to use Monto Vetiver grass ephemeral wetland for the disposal area. The Vetiver grass option would provide a solution with a considerably smaller land requirement than other irrigation systems, would be more cost effective and have much lower operating costs. The systems investigated were designed to provide a zero discharge to the creek.

Using EDVI computer model, with inputs vary from 400 to 700 KL day^{-1}, the model predicted a land area between 10 and 17 ha would be required for the complete disposal of the effluent inputs. A comprehensive management plan based on this preliminary result was developed, resulting in significant reduction in land area needed for the complete disposal, from 50 - 60 ha with pasture grass to 4 - 5 ha with Vetiver (Figure 24). This project was installed with a very significant saving in construction and maintenance costs relative to all other options.

Results to date have exceeded expectations. Only 15 months after planting, 4 ha of Vetiver has totally absorbed between 500 and 600 KL of effluent a day. Ground water monitoring showed that practically no leaching occurred during dry periods and very little during wet periods and nutrient levels in these samples were well below the license limits. Ground water monitoring is continuing.

Figure 24. Satellite image of Boonah sewage effluent treatment plant assisted with Vetiver grass technology (top), the overview of 4 ha Vetiver planting site at 12 (middle) and 18 (bottom) month old.

5.1.3.3. Catchment scale application

Citarum River Basin in Indonesia is the main source of water supply for the capital city (Jakarta) and the fourth biggest city of Indonesia (Bandung). The Basin is also used for crop irrigation and industries on the north Java plain, and for communities within the basin. However, Citarum River is known as the most polluted river in Asia. The cause of pollution is, in addition to industrial wastes both solid and liquid, indiscriminate disposal of trash and uncontrolled disposal of sewage effluent and landfill leachate to the river, using it as an open sewer (Figure 25). The Asian Development Bank (ADB), involved in a number of strategic project activities to improve the quality and performance of Citarum River Basin, has recently initiated a project using Vetiver to improve the water quality of Citarum River.

Figure 25. Citarum River is in crisis, choked by the domestic waste of nine million people and thick with the cast-off from hundreds of factories.

The strategy for pollution control of the Citarum River using Vetiver has consisted of two parts, which were implemented simultaneously:

1. Planting Vetiver on the river banks and irrigated with polluted river water (Figure 26).

Figure 26. Vetiver growth at 6 (left) and 12 months after cultivation (right)

2. Treatment of sewage effluent from communal latrine (Figure 27 and 28).

Figure 27. Vetiver planting to dispose of sewage effluent discharged from the communal latrine.

Figure 28. Vetiver growth at 2 (left) and 6 months after planting. (right)

Due to the continuous discharge of pollutants from industries along the river, the quantitative assessment of the river water cannot be made, but the qualitative observation to date indicated that the water quality has been improved, such as the infestation of blue-green algae has been substantially reduced and fish has returned to some sections of the river (Truong and Booth, 2010).

5.2. Disposal of industrial wastewater

5.2.1. Wastewater disposal from a gelatin factory and a beef abattoir

The disposal of industrial wastewater in Australia is subjected to the strict environmental guidelines enforced by the Environmental Protection Authority (EPA). The most common method of treating industrial wastewater in Queensland, Australia is by land irrigation, which is presently based on tropical and subtropical pasture plants. However, with limited land area available for irrigation, these plants are not efficient enough to sustainably dispose of all the effluent produced by the industries. The existing system used pasture species that could not meet the new EPA standards. Therefore, to comply with the new standards, most industries are now under strong pressure to upgrade their treatment processes by adopting Vetiver grass as a sustainable means of disposing wastewater (Smeal et al, 2003). The application of MEDLI model to determine land area required for cultivation of Vetiver for wastewater disposal is a practicable and cost effective solution. However, the use of MEDLI was restricted to a number of tropical and subtropical crops and pastures, except Vetiver grass, in Australia. To apply Vetiver grass in this model, therefore, it has to be calibrated first. A series of pot experiments and field trials were conducted over two year period to collect all vital information related to Vetiver calibration at Gelita Australia food processing factory and Teys Bros beef abattoir (Smeal, et al. 2003). From these results, Vetiver has been adopted and successfully used to treat/dispose of effluent generated by the factory and the abattoir.

Gelita Australia, a gelatin factory in Queensland, Australia, which extracts gelatin from cattle hide using chemical processes involving strong acids, lime and hydroxides. The effluent from the processing plant, 1.3 ML day^{-1}, is highly saline (average 600 mS cm^{-1}), alkaline and has a high organic matter content. The MEDLI computer model output based on an assumed maximum annual effluent output of 584 ML, N concentration of 300 mg L^{-1} and 121 ha available for irrigation showed that Vetiver requires the least land for sustainable irrigation in both N and effluent volume amongst the three grasses (Table 8). A reduction from 130 to 80 ha for the treatment of the effluent would result in significant cost savings of the factory (Truong and Smeal, 2003). In practice, an area of 22.5 ha was cultivated with Vetiver grass for disposing of about 48 ML monthly (Figure 29).

Table 8. Land area required by three grasses for irrigation and N disposal.

Plants	Land needed for irrigation (ha)	Land needed for N disposal (ha)
Vetiver	80	70
Kikuyu	114	83
Rhodes	130	130

Figure 29. Vetiver planted for wastewater disposal at Gelita Australia (left) and six months after planting (right).

TEYS Bros, a beef abattoir in Queensland, processes about 210 000 cattle per year for both domestic consumption and export. Effluent generated by the abattoir, approximately 1.7 ML day^{-1}, with total N of 170 mg L^{-1} and 32 mg L^{-1} of total P. Teys Bros has an area of 42.3 ha that can be used for wastewater treatment by irrigation. Effluent was either spray or surface irrigated onto Kikuyu pasture at various sites around the property. By using the MEDLI model, the simulated outcomes predicted that approximately 1.24 and 0.8 ML day^{-1} of effluent can be sustainably irrigated 42.3 ha of available land cultivated with Vetiver grass and Kikuyu grass pasture, respectively. The above result indicated that

Vetiver planting would provide an improvement of 55% over Kikuyu in planting area (Truong and Smeal, 2003). A field trial was conducted at Teys Bros to investigate the effectiveness of Vetiver in treating wastewater; the result of this study is presented in Table 9.

Table 9. Effectiveness of Vetiver planting on quality of effluent seepage at Teys Bros abattoir.

Analytes	Inlet	Outlet (mean levels in monitoring bores down slope from inlet)	
		20 m	50 m
pH	8.0	6.5	6.3
EC (μS cm^{-1})	2200	1500	1600
Total Kjel. N (mg L^{-1})	170	11.0	10.0
Total N (mg L^{-1})	170	17.5	10.6
Total P (mg L^{-1})	32	3.4	1.5

5.2.2. Wastewater from intensive animal farm

In China, the disposal of wastewater from intensive animal farms is one of the biggest problems in densely populated areas. China is the largest pig raising country in the world. In 1998, Guangdong Province had more than 1600 pig farms with more than 130 farms producing over 10,000 commercial pigs each year. These large piggeries produce 100 - 150 ton of wastewater each day, which included pig manure collected from slatted floors, containing high nutrient loads. Nutrients and heavy metals from pig farm are key sources of water pollution. Wastewater from pig farm contains very high N and P and also Cu and Zn, which are used as growth promoters in the feeds. Wetlands are considered to be the most efficient means of reducing both the volume and high nutrient loads of the piggery effluent. To determine the most suitable plants for the wetland system, Vetiver grass was selected along with other 11 species in this program. The best species are Vetiver, *Cyperus alternifolius* and *Cyperus exaltatus*. However, further testing showed that *Cyperus exaltatus* wilted and became dormant during autumn and did not rejuvenate until next spring. Full year growth is required for effective wastewater treatment. Therefore,

Vetiver and *Cyperus alternifolius* were the only two plants suitable for wetland treatment of piggery effluent (Liao et al., 2003). Results from a trial using Vetiver treatment showed that Vetiver had a very strong purifying ability. Its ratio of uptake and purification of Cu and Zn was >90%; As and N > 75%; Pb was between 30 - 71% and P was between 15 - 58%. The purifying effects of Vetiver to heavy metals, and N and P from a pig farm were ranked as Zn > Cu > As > N > Pb > Hg > P (Liao et al, 2003).

5.2.3. *Wastewater from a seafood processing factory*

In the Mekong Delta, Vietnam a demonstration trial was set up at a seafood processing factory to determine the treatment time required to retain effluent in the Vetiver field [Hydraulic Retention Time (HRT)] to reduce nitrate and phosphate concentrations in effluent to acceptable levels (Figure 30). The experiment started when plants were 3 months old. Water samples were taken for analysis at 24-hour interval for 3 days. Analytical results showed that total N content in wastewater (4.79 mg L^{-1}) was reduced by 88% and 91% after 48 and 72 hours of treatment, respectively. The total P (0.72 mg L^{-1}) was reduced by to 80% and 82% after 48 and 72 hours of treatment, respectively. The amount of total N and P removed in 48 and 72-hour treatments were not significantly different (Danh et al., 2006).

Figure 30. Vetiver planting at a seafood-processing factory in the Mekong Delta, Vietnam.

5.2.4. Wastewater from a small paper factory

In North Vietnam, wastewater discharged from a small paper factory at Bac Ninh province and a small nitrogen fertilizer factory at Bac Giang province is as highly polluted with nutrients and chemicals. The factories release their wastewater directly into a small river in the Red River Delta. Installed at both sites of the storage ponds, Vetiver became well established after two months. In general, Vetiver at the paper factory at Bac Ninh is in good shape, except for a few sections next to the polluted water, where it shows symptoms of toxicity. On the other hand, despite the highly polluted conditions, Vetiver is established and growing well at the nitrogen fertilizer factory at Bac Giang. Excellent growth has been recorded for this site under semi-wetland conditions, where Vetiver is expected to reduce pollutant levels significantly (Figure 31).

Figure 31. Vetiver at Bac Ninh (left) and Bac Giang (right).

5.2.5. Wastewater from a tapioca flour mill factory

In Thailand, three ecotypes of Vetiver (Monto, Sura Thani and Songkhla 2003) were used to treat wastewaters from a tapioca flour-mill factory. Two systems of treatment were employed, namely: (i) holding wastewater in a Vetiver wetland for two weeks and then drained off, and (ii) holding wastewater in a Vetiver wetland for one week and drain it off continuously for a total of 3 weeks. It was found that in both systems, Monto ecotype had the highest growth of shoot, root, and biomass, and was able to absorb highest levels of P, K, Mn and Cu in the shoot and root, Mg, Ca and Fe in the root, and Zn and N in the shoot.

Surat Thani ecotype could absorb highest levels of Mg in the shoot and Zn in the root, while Songkhla ecotype could absorb highest levels of Ca, Fe in the shoot, and N in the root maximally (Chomchalow, 2006).

5.2.6. Phenol contaminated water from illegal dumping of industrial waste

Over the last few years, illegal dumping of industrial waste and wastewater has become a major environmental problem in Thailand. At least 6 illegal dumping sites of phenol were discovered in Nong-Nea subdistrict, Phanom Sarakham district, Chachoengsao province. The pond of Mr Manus Sawasdee, a Nong-Nea resident who is a victim of illegal dumping, was illegal discharged with industrial wastewater with high concentration of phenol (500 mg l^{-1} at the beginning of the incident), and other hazardous organic substances, such as petroleum hydrocarbons, formaldehyde, as well as metals, namely arsenic, chromium, copper, lead, and nickel. As Mr Manus's pond is at high elevation, the runoff after raining from this pond would carry phenol and other hazardous substances along Tad Noi creek to lower waterways (Phenrat et al., 2015).

The first large-scale application of vetiver grass for treatment of phenol contaminated water was performed by cultivating 0.12 million vetiver grass bare roots to create 1.2-kilometer vetiver hedgerows along Tad Noi Creek from August 28–29, 2014. Similarly, a field-scale treatment of illegally dumped wastewater in a 768-cubic meter pond of Mr Manus using vetiver grass grown on 45 floating bamboo platforms was performed on December 5, 2014. The preliminary results of these projects were very promising in term of phenol, total petroleum hydrocarbon and COD removal. Such environmental restoration projects can be a model for the more than 50 communities recently affected by illegal dumping in Thailand (Phenrat et al., 2015) (Figure 32).

Figure 32. Farm water supply pond infected by Phenol and Vetiver pontoons treatment.

5.2.7. Wastewater from oil processing factory

In Colombia, Ecopetrol company set up a pilot-scale test to investigate the potential of Vetiver to remove fats, oils and suspended solids from wastewater generated during the oil production process. Vetiver was grown on the floating platform of an artificial wetland with dimensions of 6 m x 2 m x 1 m, water level kept at 0.6 m and water flow rate of 0.24 L second^{-1}. Preliminary results obtained in the period of 32 to 49 days after cultivation indicated that Vetiver on floating platform could remove 73 - 100% of fats and oils and 29 – 75% of suspended solids with the input range of 0.33 – 5.23 and 1.7 – 18 mg L^{-1}, respectively. The study is still continuing and expected to test the behaviour of the system and removal rates for other physical and chemical parameters. Up to date, the results are satisfactory for the treatment of waters associated with the production of oil and it is the first experimental practice using Vetiver wetland in Colombia case (Triana et al 2010).

5.2.8. Wastewater from a palm oil mill factory

Vetiver System Technology has been recently investigated in Malaysia to treat palm oil mill effluent (POME) in an attempt to reduce biochemical oxygen demand (BOD) and chemical oxygen demand (COD). POME is the product of the extraction and purification of palm oil processes that is characterized by high BOD$_3$ (350-400 mg L^{-1}) and COD (790-810 mg L^{-1}) (Darajeh et al., 2014). In this study, two different concentrations of POME (high: undiluted POME, low: volume ratio of 1 POME and 9 water) were treated with Vetiver plants for 2 weeks. The results showed that Vetiver was able to reduce the

BOD up to 90% in low concentration POME and 60% in high concentration POME, while control (without plant) was able only to reduce 15% of BOD. The COD reduction was 94% in low concentration POME and 39% in high concentration POME, while control just 12% reduction.

5.2.9. Wastewater from an Aluminum manufacturer

A pilot study was set up to test the suitability of Vetiver grass in treating effluent from Aluminum Du Maroc (an Aluminum manufacturer in Tangier, Morocco) which is heavily contaminated with Al and heavy metals (Etienne Richards, pers.com.). Following neutralization and decantation of suspended Al salt, the effluent was discharged to a Vetiver sand bed with a capacity to filter 500 L day^{-1} (6 times a day at 85 L each). Ten weeks after planting, although not fully mature, the results below show the effectiveness of Vetiver in purifying this highly contaminated effluent **(E. Richard, Kepwater, Lourdes, France, Pers.com.):**

- COD and BOD levels reduced by 98%
- Suspended solid levels reduced 99%
- Nitrogen level reduced by 95%
- Phosphorus level reduced by 97%
- Heavy metal levels reduced by an impressive 99.99% (Figures 33-34).

Figure 33. Wetland unit at planting; and Vetiver 6 week's growth during testing period.

Figure 34. Effluent quality before and after Vetiver treatment.

Following these excellent results, the company would install a treatment plant of 750 m^2 surface to treat 200 m^3 of effluent a day. This will give the company the recycling of at least 75% of total water use of this factory.

5.2.10. *Wastewater from a fertilizer company, quarry industry and a public refuse dumpsite*

The potential of Vetiver grass (*Chrysopogon zizanioides*) in the treatment of contaminated industrial wastewater has been recently investigated by Oku et al. (2015). This study was conducted in Eastern Nigeria using three sources of effluents that were generated from a fertilizer blending company, quarry industry and a public untreated refuse dumpsite. In general, leachate effluents were high in the concentration of BOD, COD, nitrate, phosphate and quite low in the concentration of lead, arsenic, zinc, iron, cadmium, mercury, nickel, copper. Vetiver grass was hydroponically grown for 10 weeks under full sunlight to allow the vetiver roots and shoots to fully establish before the experiment started. Vetiver grass was transferred to the different effluents, then the effluents were analyzed at 2, 4 and 6 days after treatment. The results of the experiment are summarized in Table 10. Vetiver significantly reduced contaminant concentrations in leachate of different sources over time. In addition, Vetiver was demonstrated to have a neutralizing effect that could adjust pH of different effluents to the neutral pH value. It can be concluded that Vetiver grass is very effective in treatment a range of contaminants in wastewater.

5.2.11. Wastewater from a mixture of laboratory and sewage source

Tran et al. (2015) investigated Vetiver grass phytoremediation capacity in treating three groups of pollutants in wastewater containing organic matters, heavy metals and aromatic compounds. Sewage effluent (SE) was firstly dilluted with tapwater with a volume ratio of 1:1, and fed to two wetland systems, mini horizontal subsurface flow (HSSF) and floating raft (FR) wetlands, for 8 weeks. Then laboratory wastewater (LW) was mixed with sewage effluent a volume ratio of 1:1, the mixture was added into two systems. Hydraulic retention time (HRT) in both systems were controlled as long as 12 hours. Quality parameters of sewage effluent and laboratory wastewater are presented in Table 11.

HSSF results revealed that even with the presence of heavy metals and aromatic compounds, Vetiver presented reasonable removal efficiencies of about 62%, 68.6%, and 58.3% for BOD, total N, and total P removal, respectively. For Vetiver roots, in term of heavy metals, had an impressive removal efficiencies of 99.2, 95.8, 96.2, and 96.7% of Cr^{+6} (in $K_2Cr_2O_7$), Mn^{2+} ($MnSO_4$), Fe^{2+} ($FeSO_4$), and Cu^{2+} ($CuSO_4$), respectively. For aromatic compounds, the wetland is responsible for 96.8 % and almost 100% of correspondingly phenol and benzene removal efficiencies.

In FR wetland, the outcomes show similar tendencies in treatment as in HSSF wetland. Vetiver grass, mainly responsible for organic matters and nutrients removal, presented slightly lower removal efficiencies than those in HSSF wetland. The average values of removal efficiencies were 59%, 63.5%, and 53.0% for BOD, TN, and TP removal, respectively. For heavy metals of Cr^{+6} (in $K_2Cr_2O_7$), Mn^{2+} ($MnSO_4$), Fe^{2+} ($FeSO_4$), and Cu^{2+} ($CuSO_4$) Vetiver root were found removing less than in HSSF wetland, with average removal efficiencies values of 92.4, 85.1, 91.8, and 91.5%, respectively.

Table 10. The removal of contaminants from public untreated dumpsite, fertilizer companyand quarry site effluent in Eastern Nigeria by Vetiver grass (*Chrysopogon zizanioides*). Source Oku et al., (2015).

Parameter/ contaminants	Public untreated dumsite effluent				Fertilizer company effluent				Quarry site effluent			
	Level of contaminants after certain days of Vetiver treatment (mg l^{-1})				*Level of contaminants after certain days of Vetiver treatment (mg l^{-1})*				*Level of contaminants after certain days of Vetiver treatment (mg l^{-1})*			
	0	2	4	6	0	2	4	6	0	2	4	6
pH	5.8	6.7	7.3	7.3	6.3	6.5	6.8	7.5	12.8	12.4	8.3	7.2
BOB	153	67.8	50.5	50	41.6	19.6	14.5	11.3	124.3	61.2	54.8	50.5
COD	151.8	68.5	52.1	47.8	29.8	16	13.3	10.6	119.8	56.5	52.5	50.8
Nitrate	115.6	51.6	47.6	42.9	122.2	58.7	28.4	7.4	120.8	53.8	18.1	5.9
Phosphate	92.9	52.7	41	40.7	55.1	36.5	15.3	12.1	64.7	40.7	10.7	4.6
Cyanide	1.02	0.71	0.09	0.06	nd	nd	nd	nd	Nd	Nd	nd	nd
Lead	nd	Nd	Nd	nd	nd	nd	nd	nd	0.3	0.26	0.07	0.01
Zinc	0.05	Nd	Nd	nd	0.89	0.38	0.06	0.03	0.18	0.073	nd	nd
Iron	1.04	0.68	0.06	0.02	0.31	0.37	nd	nd	0.83	0.3	nd	nd
Cobalt	0.1	0.07	Nd	nd	0.09	nd	nd	nd	0.04	0.04	nd	nd
Cadmium	nd	Nd	Nd	nd	0.2	0.07	nd	nd	Nd	Nd	nd	nd
Mercury	nd	Nd	Nd	nd	Nd	nd	nd	nd	Nd	Nd	nd	nd
Manganese	0.14	0.05	0.04	0.01	0.2	0.08	0.05	0.04	0.21	0.08	0.06	nd
Arsenic	0.1	0.05	0.05	nd	0.2	0.07	nd	nd	0.2	0.11	nd	nd
Nickel	nd	Nd	Nd	nd	Nd	nd	nd	nd	Nd	Nd	nd	nd
Copper	nd	Nd	Nd	nd	Nd	nd	nd	nd	Nd	Nd	nd	nd

Note: nd not detected

Table 11. Quality analysis of sewage effluent and laboratory wastewater.

Parameter/contaminants	SE	LW	Mixture
pH	6.2	5.5	6.0 ± 0.2
BOB (mg l^{-1})	420	15	220 ± 12
Total N (mg l^{-1})	65	34	55 ± 3
Total P (mg l^{-1})	10	12	11 ± 2
Cr^{+6} (mg l^{-1})	Nd	9.5	4.5 ± 0.4
Fe^{2+} (mg l^{-1})	Nd	38.5	19.8 ± 0.3
Mn^{2+} (mg l^{-1})	Nd	47.0	24.2 ± 0.6
Cu^{2+} (mg l^{-1})	Nd	35.1	17.6 ± 0.7
Benzene (mg l^{-1})	Nd	4.3	2.3 ± 0.4
Phenol (mg l^{-1})	Nd	7.8	3.8 ± 0.2

Note: nd not detected.

5.3. Disposal of landfill leachate

Disposal of landfill leachate is a major concern to all large cities, as the leachate is often highly contaminated with heavy metals, organic and inorganic pollutants. Results in Australia, Mexico, the United States and Iran showed that Vetiver growth is not affected by this highly polluted water and grows vigorously.

5.3.1. Landfill leachate disposal in Australia

Stotts Creek Landfill is a major waste depot of the Tweed Shire, New South Wales, Australia. Disposal of leachate is a major concern of the Shire as the landfill site is close to agricultural areas. An effective and low cost leachate disposal system was needed, particularly during summer high rainfall season. Following capping with topsoil, Vetiver was planted on the surface of landfill mound and irrigated with leachate from collecting ponds (Figure 35). Results to date have been excellent. In the second year, Vetiver with 3 m high forming the thick and tall walls was recorded. The growth was so vigorous that during the dry period, there was not enough leachate in the ponds to irrigate both the old and new plantings. A planting of 3.5 ha in January 2003 has effectively disposed of 4 ML a month in summer and 2 ML a month in winter (Percy and Truong, 2005).

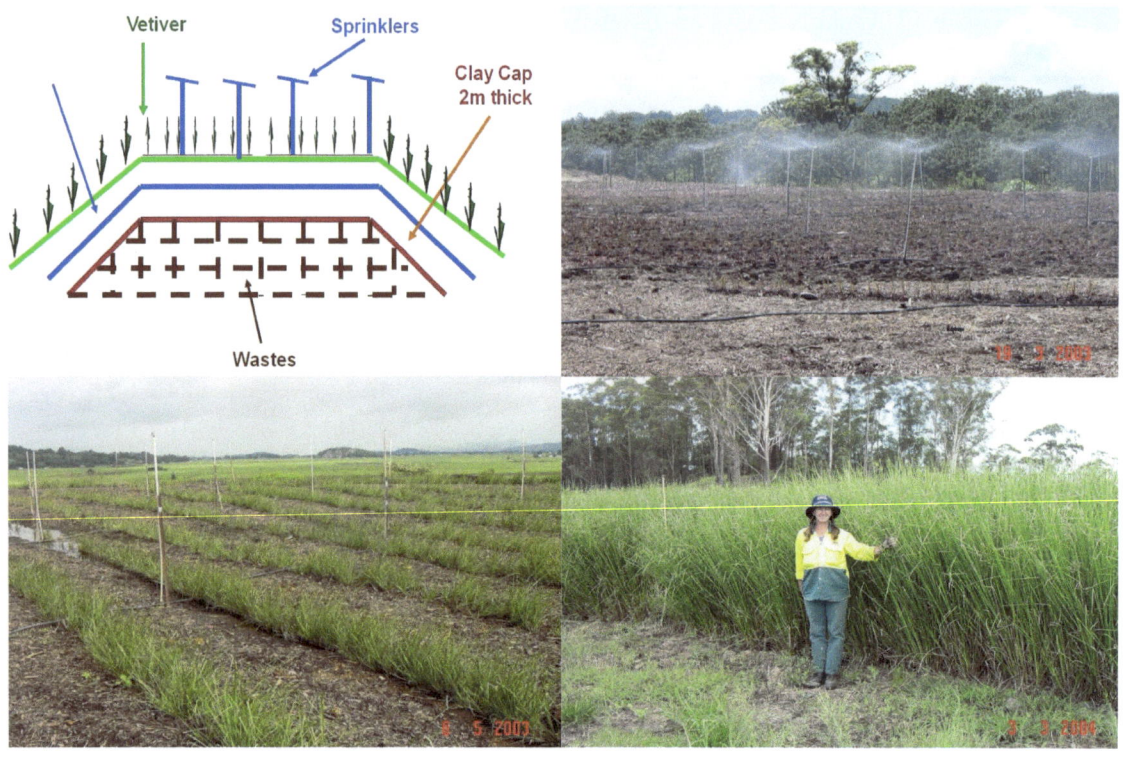

Figure 35. The Vetiver treatment at Stott Creek landfill mound: the diagrammatic cross section of the mound (top left), Vetiver irrigated every day with leachate after planting (top right), two (bottom left) and twelve (bottom right) months after planting.

5.3.2. Landfill leachate disposal in Mexico

PASA, which is the largest solid waste management company in Mexico, currently applies VPT for landfill leachate disposal, leachate seepage mitigation and for erosion control on landfill side slopes. The company has three projects underway, one each at Leon, Poza Rica, and Villahermosa. These projects have some 300,000 Vetiver plants in the ground (Figure 36). PASA is also planning for additional three Vetiver projects in Mexico, and likely in Belize. The Leon landfill has a very strong, fresh domestic and industrial leachate. In addition to the 25,000 gallons of leachate produced daily, an additional 15 million gallons are currently stored in lagoons waiting for treatment. The Poza Rica facility uses Vetiver for three main purposes: the stabilization of very steep, highly erodible slopes, on-site utilization of fresh leachate, and control of leachate outbreaks. Villahermosa is similar to Poza Rica, but the design and operation of an effective system was further complicated due to the extreme rainfall at this facility, which is located along the southern coast of the Gulf of Mexico (Truong et al, 2012).

Figure 36. Early stage of Vetiver establishment at Leon (left) and Poza Rica (right).

5.3.3. Landfill leachate disposal in Morocco

A very large new landfill complex at Oujda City, near the eastern border of Morocco and Algeria, is being built. When various options were considered for the disposal of a highly concentrated leachate, a product of combined industrial and domestic wastes, Vetiver was recommended and is being implemented now (Figure 37).

Figure 37. Fresh industrial and domestic wastes being compacted (left) and site ready for Vetiver planting (Etienne Richards, pers.com.).

5.3.4. Landfill leachate disposal in the United States

In the USA, Leggette, Brashears & Graham, Inc. has used hybrid poplar very successfully as a phytoremediation method near St. Louis and Chicago. After learning about Vetiver, the company switched to grass from trees and used Vetiver at the Republic Services Gulf Pines landfill near Biloxi, Mississippi. The phytoremediation system using VS was set up to utilize up to 14 ML year^{-1} (3 million gallons) of leachate. This was the first of its kind for large-scale project in the US and Western Hemisphere to use Vetiver grass (Truong et al., 2012). From a technical standpoint, the approach blended numerous disciplines including engineering, hydrology, microbiology, plant physiology/morphology, soil science, agronomy, chemistry, hydrology, and computer science (PLC programming and evapotranspiration modeling). With an area of 3 acre cultivated with over 50,000 Vetiver plants (Figure 38), the system has performed as designed and 100% of leachate generated has been utilized on site, well ahead of anticipated results.

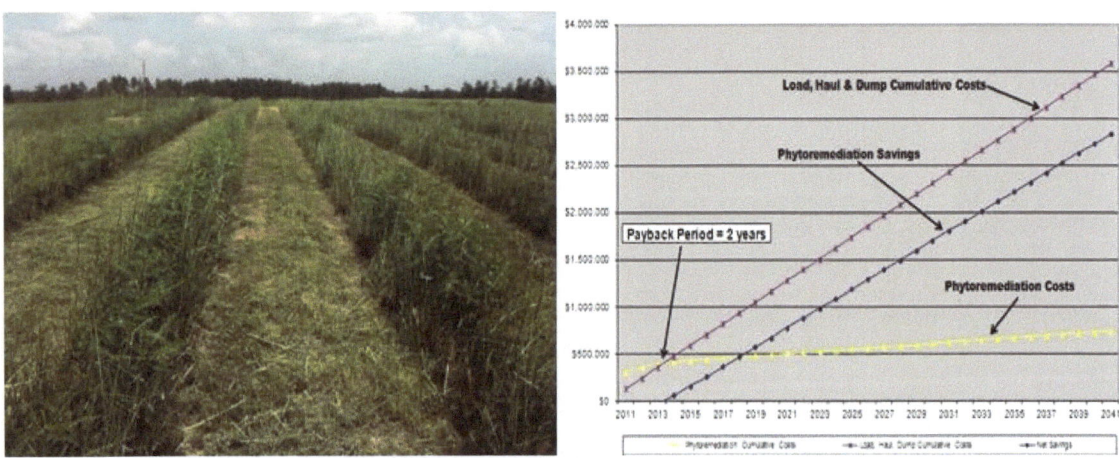

Figure 38. Vetiver growth after 7 weeks (left), the approach greatly reduces costs in an environmentally friendly manner.

The results to date have exceeded expectation in term of utilization and cost saving:
- Approximately 500,000 gal processed in first 4 months, greater than 1 million in the first year
- $150,000 transportation and disposal costs avoided in the first year
- Pre-Vetiver leachate disposal cost = $0.09 gal^{-1}
- Post-Vetiver leachate disposal cost ≤ $0.01 gal^{-1}
- Return on initial capital investment: only 2 - 3 years

- $3 million expected savings over 30 years

Due to the outstanding performance, this project was awarded a "2012 Grand Prize – Small project for Excellence in Environmental Engineering" by American Academy of Environmental Engineers (http://www.aaee.net/E32012GPSmallProjects.php).

5.3.5. Landfill leachate disposal in Iran

Vetiver has been demonstrated to survive and adapt well to the harsh conditions of landfill located in Shiraz city, one of the biggest cities in Iran (Jalalipour et al., 2015). The landfill has a total area of 40 ha. It has a semi-arid climate with mild winters and average annual precipitation of 389 mm mainly in autumn and winter. Average temperatures in the coldest and warmest months range from 6.7 to 28.2°C. Potential evapotranspiration is 5 mm per day and 1,825 mm per year. Average annual wind speed is 200 km/h at 2 m height (Master plan of Shiraz solid waste management, 2009). The landfill was estimated to generate about 120 m^3 of leachate per day in 2013. The leachate, with high organic content (BOD_5, COD), high concentration of heavy metals, ammonia, toxic compounds, bacterial contamination and unpleasant odor creates environmental and health problems, unseemly sights and adverse effects on soil and water resources. To reduce the adverse impacts of landfill leachate, green vegetation was established on 20 ha of the landfill site with fruit and fruitless trees in 2000. In addition, 780 ha of olive trees and 100 ha of forestry trees (eucalyptus, pine and cypress) were cultivated around the landfill. However, the green cover failed to remediate leachate due to dry conditions, high wind speed and toxic characteristics of leachate. Vetiver phytoremediation technology (VPT), a simple and affordable solution, has been attempted to control the quantity of landfill leachate. In a field trial to test Vetiver adaptability to Shiraz landfill environment, Vetiver was cultivated on about 45 m^2 of an open cell (fully loaded with solid wastes and covered with 3-4 m of soil). Results of this trial showed that Vetiver established and grew well and its growth was not affected by strong winds, while olives (Figure 39) and eucalyptus trees were adversely affected. The results of a greenhouse experiment showed that Vetiver can tolerate irrigation with 45% of leachate. In addition, cultivation of Vetiver in large scale offers a pleasant view as an additional advantage, as well as to become a very effective soil and water conservation measure. It can be concluded that Vetiver grass cultivation is the best option for Shiraz landfill capping after closing.

Figure 39. Olive (left) and Vetiver grown on Shiraz landfill.

5.4. Municipal landfill leachate seepage control

Vetiver grass is also very effective in controlling leachate seeping from the side slopes of a 30 years old landfill mounds in Cleveland, Queensland, Australia. This seepage was highly contaminated with Cr, Cd, Cu, Pb and Zn. It would eventually run into a nearby creek. One year after planting, Vetiver showed excellent growth that was not affected by heavy metals contamination in the leachate, and completely stopped the leachate seepage (Figure 40).

Similarly, Vetiver has been successfully applied to control seeping leachate from a landfill in Guangdong province, China as well as to stabilize its dam wall. The landfill was built in a valley with a surface area of over 23 ha and currently takes 2500 tons of waste a day from Guangzhou City. Two earthen walls of the landfill were built across the valley floor with semi-weathered rocks and clay but they were not properly designed and well-built as normal dam walls. The city garbage was then dumped and compressed into the space between the two walls. As the garbage reached a few meters high, the surface was covered with earth followed by the heavy geo-membrane to cover the whole surface. Once the space was completely filled with wastes, the two walls were raised to take more garbage. The walls are now 75 m high and 100 m long, and under very high pressure caused by both the large amount of garbage and heavy machinery working on the surface layer. Consequently, a large quantity of leachate seeped through the wall causing slippage and erosion in rainy season. Previous efforts to stabilize the wall by using both native and imported plant specieshave failed because of the toxic nature of the leachate. Vetiver was employed in an attempt to stabilize the dam wall and to reduce seeping leachate. Although the soil conditions were extremely hostile, such as crushed weathered rock, highly compacted and very low in nutrients, Vetiver established well and not only succeeded in stabilizing the dam wall, but also dried up leachate seepage (Figure 41). Vetiver could grow well on the edge of highly toxic leachate pools; both native and introduced plants were killed (Percy and Truong, 2003).

Figure 40. Leachate after rain on the side slope of the old landfill mound (top), within a year Vetiver completely stopped the leachate seepage (bottom).

Figure 41. Vetiver planting on the slope of this old landfill mound in Guangzhou (left), leachate seepage was completely stopped one year after planting (right).

5.5. Reducing toxic elements in irrigation water

Ugalde Smolcz and Goykoviv Cortés (2015) used Vetiver phytoremediation technology to remediate Boron contaminated water and soil for agricultural crops in Chile. The valleys of Arica Parinacota Province in Northern Chile present outstanding climatic conditions that allow crop production all year long, the province is the supplier of fresh vegetables for both central and southern Chile during winter, placing Arica´s valleys as one of the key factors in food security of the country.

However, the valleys are inserted in a desert region where salinity, boron and arsenic are in high concentrations in rivers, as well as in soil, restricting the development of most plant species. The purpose of this study was to evaluate an unconventional strategy for boron remediation in irrigation water and agricultural soil of the Lluta valley as follows:

- Different Vetiver biomasses were tested in a 3000 L pool. The efficiency remediation was 20-23% for the 5, 10, 20 and 25 kg biomass and 36% for the 15 kg biomass treatment. The efficiency removal was 98.4% for lead, 40% for arsenic and 76% for manganese. Boron level decreased by 2 mg/L.

- A field experiment was established to introduce 4 new crops and irrigated them with Vetiver treated water. Crop yields in the Azapa valley were used as control since this valley has no problems with boron and salinity. Yields for corn were high, with 1 corncob per plant. All corn was 100% extra quality. This is a very significant result since previously sweet corn cannot be grown in this valley. Lettuce had a yield of 4 boxes (12-14 lettuces) every 10 linear meters. For melon the average was 3 melons per plant of 2nd class. For Cristal chili pepper, yield reached 70-80 fruits per plant of first class.

- A soil test was established in pots with 6 replications that were irrigated with different boron concentrations. The treatments consisted in T1; 1 mg/L, T2; 20 mg/L, T3; 50 mg/L; T4; 100 mg/L. The pots were irrigated for 3 months and samples were taken every 4 weeks for soil and leaves analyses. Efficiency percentage at 3rd month were: T1; 66.3%, T2; 91%, T3; 95%, T4; 96.5%.

It was concluded that Vetiver Phytoremediation Technology is a technology capable of remediating boron toxicity, allowing the introduction of new crops and improvement of crop yields in valleys of Arica Parinacota Province (Figure 42).

Figure 42. Irrigation pool with Vetiver floats, left. Extra quality sweet corn, right

VI. PREVENTION, TREATMENT AND REHABILITATION OF MINING WASTES AND CONTAMINATED LANDS

There have been increasing concerns in Australia and worldwide on the contamination of the environment by by-products of rural, industrial and mining industries. The majority of these contaminants are high levels of heavy metals which can affect flora, fauna and humans living in the areas, in the vicinity or downstream of the contaminated sites. Table 12 shows the maximum levels of heavy metals tolerated by environmental and health authorities in Australia and New Zealand.

Concerns about the spreading of these contaminants have resulted in strict guidelines being set to prevent the increasing concentrations of heavy metal pollutants. In some cases industrial and mining projects have been stopped until appropriate methods of decontamination or rehabilitation have been implemented at the source.

Methods used in these situations have been to treat the contaminants chemically, burying or to remove them from the site. These methods are expensive and at times impossible to carry out, as the volume of contaminated material is very large, examples are gold and coal mine tailings.

If these wastes cannot be economically treated or removed, off-site contamination must be prevented. Wind and water erosion and leaching are often the causes of off-site contamination. An effective erosion and sediment control program can be used to rehabilitate such sites. Vegetative methods are the most practical and economical, however, revegetation of these sites is often difficult and slow due to the hostile growing conditions including high concentrations of heavy metals and pollutants, extreme pH, low nutrients, high salinity, low or high moisture soil content and coarse or fine soil texture.

In term of environmental protection, the most significant breakthroughs in the last 20 years are firstly research leading to the establishment of benchmark tolerance levels of Vetiver grass to adverse soil conditions and secondly its tolerance to heavy metal toxicities. These have opened up a new field of application for VS: treatment of mining wastes and contaminated lands by rehabilitation.

Table 12. Thresholds for contaminants in soils (ANZ, 1992).

Heavy metals	Threshold (mg kg^{-1})	
	Environmental*	Health*
Antimony (Sb)	20	-
Arsenic (As)	20	100
Cadmium (Cd)	3	20
Chromium (Cr)	50	-
Copper (Cu)	60	-
Lead (Pb)	300	300
Manganese (Mn)	500	-
Mercury (Hg)	1	-
Nickel (Ni)	60	-
Tin (Sn)	50	-
Zinc (Zn)	200	-

*Maximum levels permitted, above which investigations are required.

Soil disturbance on mine-sites inevitably leads to erosion and to the transport of sand, silt and clay particles in runoff water. This sediment load has the potential to cause environmental harm further downstream. Trapping and retaining the sediment on the mining lease is a legal requirement.

The underlying principle behind sediment control is to reduce the velocity of runoff water. This causes suspended soil particles to settle out. The larger course sand particles settle first, followed by fine sand, silt and then clay. Some clay particles may stay in suspension and can only be precipitated using chemicals such as gypsum.

Conventional control measures to reduce the velocity of storm water include engineered structures such as diversion drains and silt traps (sometimes called silt or sediment retention ponds), hay bales and silt mesh fences are used for short-term silt traps. On some mine sites, 'dirty' storm-water is filtered by channeling it through a wetland. A series of small sediment traps is more effective than one big one. The separation of 'clean' and 'dirty' water is an important principle.

Vetiver grass can be used in almost any situation where erosion control or sediment control is required. Several hedges can be planted across a gully at strategic points.

Further down the slope, the effectiveness of a conventional silt trap can be increased by planting Vetiver hedges across the spillway. A double row of Vetiver is more effective than a single row.

Followings are case studies of applying Vetiver grass for reclamation of mining wastes and contaminated lands around the world.

6.1. Gold mine

Vetiver was successfully used for a large-scale application to control dust storm and wind erosion on a 300 ha dam of fresh gold tailings at Kidston gold mine, Queensland, Australia. Typically, fresh gold tailings are alkaline (pH 8-9), low in plant nutrients and very high in free sulphate (830 mg kg^{-1}), sodium and total sulphur (1-4%). When dry the finely ground tailings material can be easily blown away by wind storms if not protected by a surface cover (Figure 43). As gold tailings are often contaminated with heavy metals, wind erosion control is a very important factor in stopping the contamination of the surrounding environment. The usual method of wind erosion control in Australia is by establishing a vegetative cover, but due to the highly hostile nature of the tailings, revegetation is very difficult and often failed when native species are used. The short term solution to the problem is to plant a cover crop such as millet or sorghum with protecting fences to promote crop establishment (Figure 44), however this approach does not last very long (Figure 45). Vetiver can offer a long term solution by planting into rows at spacing of 10 m to 20 m to reduce wind velocity and at the same time provide a less hostile environment, such as shading and moisture conservation, for crop establishment at the beginning and local native species to establish voluntarily later (Figure 46). Vetiver established and grew very well on these tailings without fertilizers, but growth was improved by the application of 500 kg ha^{-1} of DAP.

Figure 43. A typical large fresh tailings pond of gold mine wastes (left), strong wind causing a dust storm which contains high levels of heavy metals (right).

Figure 44. A conventional measure included planting a surface cover crop and building fences to control wind erosion promoting crop establishment.

Figure 45. Despite its very solid construction, these rigid and expensive fences are also vulnerable to high wind velocity.

Figure 46. The flexible Vetiver hedges provided a low cost and permanent wind barrier unaffected by strong winds, provided excellent protection for crop establishment (2 years after planting), ten years after planting, no fertilizers and heavy grazing.

Similarly, in field trials Vetiver showed good establishment on old gold mine tailings at Kidston gold mine that are commonly characterized by extreme acid (pH 2.5 – 3.5), high in heavy metals and low in nutrients. These tailings are often the source of contaminants for both above-ground and under-ground to local environment. Table 11 shows the heavy metal profile of gold mine tailings in Australia. At these levels some of these metals are toxic to plant growth and also exceed the environmental investigation thresholds (ANZ, 1992). Furthermore, the bare soil surface is highly erodible (Figure 47). Therefore, revegetation of these tailings is very difficult and often very expensive. The field trials of Vetiver were conducted on two old (8 year) gold tailings sites. One is typified by a soft surface and the other with a hard crusty layer. The soft top site had a pH of 3.6, sulphate at 0.37% and total sulfur at 1.31%. The hard top site had a pH of 2.7, sulphate at 0.85% and total sulfur at 3.75% and both sites were low in plant nutrients. Results from both sites indicated that when adequately supplied with nitrogen and phosphorus fertilizers (300 kg ha^{-1} of DAP) excellent growth of Vetiver was obtained on the soft top site without any liming. But the addition of 5 t ha^{-1} of agricultural lime significantly improved Vetiver growth. Although Vetiver survived without liming on the hard top site, the addition of lime (20 t ha^{-1}) and fertilizer (500 kg ha^{-1} of DAP) improved Vetiver growth greatly (Figure 47).

Table 13. Heavy metal contents of representative gold mine tailings in Australia

Heavy metals	Total contents (mg Kg^{-1})	Threshold levels (mg Kg^{-1})
Arsenic	1120	20
Chromium	55	50
Copper	156	60
Manganese	2 000	500
Lead	353	300
Strontium	335	Not available
Zinc	283	200

Figure 47. Highly erodible bare soil surface of old gold tailings (left) and good establishment of Vetiver on the hard top site with amendment of lime and fertilizers (right).

In late 2010, a new gold mine development at Toka Tindung gold mine in North Sulawesi, Indonesia adopted VST for mitigation of the environmental problems before completing the entire infrastructure. In January 2011, about 100,000 Vetiver slips were planted in the most vulnerable locations, following recommendations and designs by Indonesia Vetiver Network (Figure 48). The mining company also plans to involve the local communities by providing them with Vetiver awareness training and to supply Vetiver for local nurseries that will sell to the mining company to supply their ongoing needs.

Figure 48. Design and application of Vetiver grass technology throughout extensive infrastructure and drainage. Source: www.vetiver.org.

Vetiver system has been successfully applied to rehabilitate gold mine tailings dam in South Africa (Figure 49 and 50). Recently, Vetiver has also been applied at Anglo America gold mine in Guinea, West Africa (Figure 51).

Figure 49. Gold mine tailings dam before (left) and after three month of Vetiver planting (right) (Tony Tantum pers.com.).

Figure 50. The same gold mine tailings dam after three years of Vetiver cultivation (Tony Tantum pers.com.).

Figure 51. Vetiver application at Anglo America gold mine in Guinea, West Africa. Source: Noffke, 2013.

6.2. Coal mine

6.2.1. Overburden

Open-cut coal mining is usually applied where coal deposits are close to surface, and results in the damage of the natural ecosystem. During open-cut mining, the overlying soil and fragmented rock are removed and left over on the land in the form of overburden dumps. These occupy large amount of land, which loses its original use and generally gets soil qualities degraded (Barapanda et al., 2001). As the dump materials are generally loose, fine particles from it become highly prone to blowing by wind. It has been found that overburden dump top materials are usually deficient in major nutrients (Rai et al., 2011). Hence, most of the overburden dumps do not support revegetation.

Vetiver can grow well on both fresh and old coal mine overburdens, leading to successful revegetation of these sites. Vetiver was planted on contour line to conserve soil moisture and stabilize loose surface materials of highly erodible alkaline and sodic overburden of open-cut coal mine in central Queensland (Figure 52). Vetiver grew well and promoted the establishment of native plants after 18 months of cultivation, particularly the surface of this overburden was almost completely covered by Vetiver, native trees and grasses after 9 years (Figure 53). Similarly, Vetiver has successfully established on an old coal mine overburden that remained barren after 50 years in central Queensland, and stabilized its spoil dump with slope higher than 45° to stop gully erosion and trap sediment (Figure 54). Subsequently, Vetiver has promoted the establishment of other sown and native pasture species.

Figure 52. Vetiver grew on contours of fresh coal mine overburden.

Figure 53. The site surface covered by Vetiver, native grasses and trees after 18 months (left) and 9 (right) years of cultivation.

Figure 54. Old coal mine before (left) and after 1 year of planting Vetiver (right).

6.2.2. Tailings

Vetiver was demonstrated to thrive on a coal mine tailings in a field trial in Queensland, Australia (Figure 55). The objective of this investigation was to select the most suitable species for the rehabilitation of a dam of coal mine tailings with the surface area of 23 ha and capacity of 3.5 million cubic meters. This site was characterized by high salinity, sodicity and extremely low content of nitrogen and phosphorous. In addition, it also contained high level of soluble sulfur, magnesium, calcium as well as plant available copper, zinc and iron. Five salt tolerant species were used in the study, namely Vetiver, marine couch (*Sporobolus virginicus*), common reed grass (*Phragmites australis*), cumbungi (*Typha domingensi*) and *Sarcocornia* spp. After 210 days of cultivation, complete mortality was recorded for all species except Vetiver and marine couch. The survival of Vetiver was significantly increased by mulching, however fertilizer application had no effect. The combination of mulching and fertilizers increased growth of Vetiver by 2 t ha^{-1} which was almost 10 times higher than that of marine couch (Radloff et al., 1995). The results confirm the findings from glass house trials.

Figure 55. Vetiver survived after 210 days of cultivation.

Recently, Vetiver grass technology has been applied at coal mines in South Kalimantan, Indonesia with three distinct purposes.

1. Rehabilitation of coal mine tailings slopes and embankment: the tailings are characterized by low fertility, sandy texture with the slope of 27%. It is too difficult for revegetation even with hydroseeding technology. The preliminary results after four months of cultivation showed that Vetiver grew well on the slope

of this tailings (Figure 56), particularly it promoted the growth of other plant species (Figure 57).

2. Improvements of run-off water quality: Vetiver hedges act as bio-filters that slow down the water flow rate and trap sediments leading to the release of run-off/waste water with better quality into the surrounding environment (Figure 58).

3. Stabilization of channel banks of waste-water disposal ditches (Figure 59).

Figure 56. Coal mine tailings (left) planted with Vetiver (right). Source: www.vetiver.org.

Figure 57. Vetiver acted as a pioneer species and very soon the growth of other plant species. Source: www.vetiver.org

Figure 58. Sediments trapped by Vetiver after only 4 months after planting. Source: www.vetiver.org

Figure 59. Vetiver planted for channel bank stabilization (along the channel) and sediment traps (cross the channel). Source: www.vetiver.org

6.3. Bentonite mine

Waste materials from bentonite mine tailings in Miles, Queensland, are highly sodic, high in sulphate and extremely low in plant nutrients (Table 14). These materials are highly erodible due to the highly dispersive characteristics of the sodic soil when wet. Revegetation on the tailings has been very difficult as sown species were often washed away by the first rain and the remains could not thrive under these harsh conditions.

Several field trials were conducted to investigate the establishment of Vetiver grass on one of the major disturbed areas of this mine and the effectiveness of Vetiver hedges in spreading concentrated flows and trapping sediment over major flow areas, in providing a support mechanism for other plant growth and in reducing signs of visible erosion (Bevan et al., 2000). It is due to the fact that one of the main environmental concerns associated with bentonite mining is the effect of run-off water from disturbed areas to surrounding catchments, particularly with sediment being the principal transport mechanism for a range of pollutants entering water courses (Kingett et al., 1995).

With adequate supply of fertilizers and water, Vetiver successfully established on the tailings (Figure 60). The Vetiver hedge was very effective in trapping both coarse and fine sediment, reduced visible signs of erosion and conserved soil moisture. Importantly, the combination of these effects contributed to the improvement of seedbed conditions resulting in the establishment of indigenous species.

Table 14. Chemical analyses of bentonite mine overburden and wastes in Miles, Queensland, Australia.

Analyses	Overburden	Bentonite waste
pH	5.4	5.4
EC (mS cm^{-1})	0.18	0.14
Cl (mg kg^{-1})	135	47.4
NO$_3$-N (mg kg^{-1})	1.9	0.7
P (mg kg^{-1})	2	5
SO$_4$-S (mg kg^{-1})	66	101
Ca (meq 100^{-1}g^{-1})	0.19	0.93
Mg (meq 100^{-1}g^{-1})	4.75	6.44
Na (meq 100^{-1}g^{-1})	2.7	7.19
K (meq 100^{-1}g^{-1})	0.16	0.43
Organic matter (%)	0.45	0.35
ESP (%)	35	48

Note: EC electrical conductivity, ESP exchangeable sodium percentage.

Figure 60. Bentonite mine tailings dump with barren surface (left), after 14 months of Vetiver cultivation the growth of other species observed (right).

6.4. Bauxite mine

Vetiver is currently being used to stabilise walls of a very large dam containing residues at a bauxite mine in Gove, northern Australia. Over 12 hectares of dam wall faces were planted with Vetiver grass to help anchor the topsoil and control erosion (both rill and gully development). Furthermore, several field trials were conducted to investigate the possibility of establishing Vetiver on the highly caustic tailings (red mud, fresh and old residue sands) that have pH level as high as 12. If successful, Vetiver can be used to revegetate these tailings *in situ* without capping its surface first with topsoil, which is not generally available and over time the capillary rise of sodium and alkalinity will degrade the topsoil cover. This will affect growth of plants that have low tolerance to sodicity and alkalinity. Preliminary results indicated that Vetiver can grow well on modified bauxite red mud and residue sands (Figure 61 and 62).

Figure 61. Vetiver at three weeks after planting with only amendment of nitrogen and phosphorous fertilizers.

Figure 62. Good establishment of Vetiver on residue sand except some extremely caustic spots.

Currently, the use of Vetiver has been incorporated into the general policy of CVG Bauxilum open-cut bauxite mine, located in Los Pijiguaos, Bolivar State, Venezuela, for mitigating the impact of mining activities on environment and the local community due to its success after three years of cultivation (Luque et al., 2006). The characteristics of soils at this mine are extremely low in plant nutrients and organic matters, variably physical conditions according to physiographic positions, in general of a high readability, and pH values of 4-5. Vetiver can grow well on such soils leading to its success in several applications. Firstly, Vetiver has effectively controlled the erosion occurred on various gradient slopes, gullies, border drains, the soil-concrete interfaces and roadside ditches located in a very erodible ground and high rainfall zone. Secondly, Vetiver barriers formed in the roadside ditches have reverted the erosion process by catching sediments and forming terraces. Finally, Vetiver barriers have reinforced lagoon dikes and also acted as sediment filters (Figure 63-65). Consequently, Vetiver has reduced the amounts of sediments released from mining operations into the surrounding water courses and promoted the establishment of other native plant species. For erosion control, a total of 26300 m of Vetiver barriers were planted from 2003 to June 2006 and another 7400 m of Vetiver barriers have been subsequently planted. The huge amount of cultivated Vetiver plants at this mining site has offered a great opportunity for the local communities, which have been affected by mining activities, in term of economic and environmental aspects. The locals have been trained to be craftsmen by using Vetiver leaves (that are frequently harvested to maintain Vetiver growth) for the production of handicrafts (Figure 66). They have been also taught to commercialize their elaborated products. These activities have contributed to the income of the locals. Particularly, Vetiver leaves have replaced the leaves of the moriche palm (*Mauritania flexuosa*), ancestrally used by different Venezuelan ethnic groups for the handcrafts preparation. The moriche palm plays a vital role in this ecosystem, however the permanent exploitation of the leaves has significantly decreased its population. In conclusion, during the past three years CVG Bauxilum has successfully adopted the VST for land rehabilitation and environmental protection to restore this open cut bauxite mining site of Venezuela, to a desirable environmentally friendly level (Luque et al., 2006).

Figure 63. Various slopes (left) stabilized by Vetiver (right). Source: Luque et al., 2006.

Figure 64. Gullies (left) stabilized by Vetiver grass (right). Source: Luque et al., 2006.

Figure 65. Vetiver planted for reinforcement of lagoon dikes (left) and 14 months after planting (right). Source: Luque et al., 2006.

Figure 66. The locals were trained to make handicrafts from Vetiver leaves. Source: Luque et al., 2006.

6.5. Copper mine

The copper mining is one of the main financial sources of Chile. However, the wastes produced by this mining industry can represent a huge source of contaminants to the environment - water, soil and air if they are not managed properly. Currently, most of the wastes are not reprocessed and reused in the production processes; hence storage is the only viable option for management. The residues are totally devoid of organic matter; very low contents of essential plant nutrients (0.1 mg kg^{-1} total nitrogen, 0.1-0.2 mg kg^{-1} total phosphorous) and very high levels of copper (2369 -2420 mg kg^{-1}). Consequently, native vegetation cover cannot establish on these surfaces for years leading to wind and water erosion that spread the contaminants into the surrounding environment. Recently, the application of Vetiver grass for rehabilitation of various mine tailings has been demonstrated to be effective, inexpensive and easy to implement. In 2005, a series of demonstration trials using Vetiver grass technology were set up at two copper mines in the central region of Chile to study: 1) whether Vetiver can grow on highly contaminated copper waste rocks and tailings dam as well as in the extreme weather conditions, namely high altitude, cold and wet winter, very hot and dry summer at Lo Aguirre mine, 2) whether Vetiver is effective in stabilizing the tailings dam wall (built with copper tailings material only) and in protecting fresh and old tailings dams as well as waste rock dump from wind and water erosion at El Soldado mine (Fonseca et al, 2006). The encouraging results from these trials were presented in the Latin American Vetiver Conference in Santiago, Chile (Arochas et al, 2010) as followings.

At Lo Aguirre mining site, after 3 months of planting about 80% of cultivated Vetiver plants survived and grew well on highly contaminated copper waste rocks and tailings dams, but some of them were grazed by rabbits and horses (Figure 67). After five years of cultivation, a poor survival rate of Vetiver recorded (15%) was mainly due to dehydration and herbivores (Figure 68). Many plants were observed with unrecoverable damage and some with complete disappearance. However, the survived plants were not noticed with any damage and showed good development with plant height of over 100 cm. There was no significant difference observed between Vetiver planted with and without topsoil. The findings provided an important evidence that Vetiver can survive strong aridity and intense cold in 4 years since its planting. The dry plants shown in Figure 68 are the dormant form of Vetiver in winter and they will regrow upon the arrival of spring season. The study also confirmed that Vetiver can grow well at an elevated place with the altitude of 3,500 m.

Figure 67. Vetiver establishment on copper waste rocks (left) and Vetiver grazed by herbivores (right) after 3 months of cultivation. Source: Arochas et al, 2010.

Figure 68. Vetiver growth (left) and grazed (right) by herbivores after 5 years of establishment. Source: Arochas et al, 2010.

At El Soldado mining site, all Vetiver plants survived on sand tailings and reached a height of about 35 cm after 2 months of cultivation (Figure 69). However, after 5 years of establishment without irrigation (except the first 7 months after planting) and fertilization, only 25% of Vetiver was observed to be survive (Figure 70) with the extensive root system of about 85 cm depth (Figure 71). It may be concluded that majority of Vetiver cannot adapt to the hostile conditions of new planting site without irrigation and fertilization in a long period.

Figure 69. The growth of Vetiver on copper sand tailings after 2 months of planting. Source: Arochas et al, 2010.

Figure 70. Vetiver growth after 5 years of cultivation. Source: Arochas et al, 2010.

Figure 71. Vetiver root system after 5 years of cultivation. Source: Arochas et al, 2010.

From the above results, to achieve optimum plant acclimation should have at least:
- Adding fertilizers to provide nutrients to the soil.
- Irrigation at least twice a week in summer during the first year.
- Protection against herbivores

6.6. Lead/zinc mines

Lechang lead (Pb)/ zinc (Zn) mine in the northern part of Guangdong Province, PR China, employs the underground mining operation that covers an area of 1.5 km^2, and produces approximately 30,000 tonnes of tailings annually with a dumping area of 60,000 m^2 (Shu and Xial., 2003). The climate of this mine is subtropical and the annual rainfall is about 1,500 mm. The Pb/Zn tailings contained high concentrations of heavy metals (total Pb, Zn, Cu and Cd concentrations 4164, 4377, 35 and 32 mg kg^{-1} respectively), and low contents of major nutrient elements (N, P, and K) and organic matters. The toxicity of heavy metals and deficiency of major nutrients represent the main limiting factors for plant establishment on mine tailings.

A field trial was conducted to compare the growth of four grasses (*Vetiveria zizanioides*, *Paspalum notatum*, *Cynodon dactylon* and *Imperata cylindrica* var. major) on Lechang Pb/Zn mine tailings with different amendments, with an ultimate goal of screening the most useful grass and the most effective measure for revegetation of tailings (Shu and Xia, 2003). The results showed that the height and biomass of Vetiver were significantly greater than those of the other three grasses. In other words, the growth performance of Vetiver was the best among the tested species under the same amendment (Figure 72). The domestic refuse and NPK fertilizer could improve plant growth, and their combination achieved the best growth. After six months, Vetiver under the treatment of the domestic refuse and NPK fertilizer had 100% coverage and the dry yield of 2111 gm^{-2}. The metal analysis showed that the concentrations of Pb, Zn and Cu in shoots and roots of Vetiver were significantly less than those of the other three species, and the shoot/root metal concentration quotients for Pb, Zn and Cu in Vetiver were also lower than those of other three species. These results indicated that Vetiver was more suitable for phytostabilization of toxic mined lands than *P. notatum* and *C. dactylon*, which accumulated a relatively high level of metals in their shoots and roots.

Figure 72. The superior growth of Vetiver on Pb/Zn mine tailings as compared to other grasses. Source: Shu and Xia, 2003.

Another field trial was also conducted at Lechang Pb/Zn mine but at a different tailings pond to evaluate the effect of applying domestic refuse and NPK fertilizer on Vetiver growth, and to compare the growth performance and heavy metal accumulation of Vetiver and two legume species (*Sesbania rostrata* and *S. sesban*) (Shu and Xia, 2003). Biomass of Vetiver significantly increased after the application of domestic refuse, and Vetiver grew best in tailings amended with domestic refuse and NPK fertilizer (1,111 g m^{-2}). The results further indicated that domestic refuse was a useful ameliorative material for improving physio-chemical characters of the toxic tailings. Among the three plants tested, Vetiver had the highest tolerance to metal toxicities (Figure 73) and accumulated the lowest concentrations of heavy metals in the shoots among the three species. This species was considered more suitable for stabilizing mine tailings, with the danger of transferring toxic metals to grazing animals was minimal (Yang et al., 2003).

Figure 73. The growth of Vetiver and two legumes on Pb/Zn tailings.

Vetiver grass has been successfully applied to rehabilitate the contaminated land surrounding Shaoguan Pb/Zn smelting factory in the north of Guangdong province, about 50 km away from Lechang Pb/Zn mine (Shu and Xia, 2003). The dust and gas has emitted from Pb/Zn smelter-refinery processes contained high concentrations of SO_2 and heavy metals, such as Pb, Zn, Cd and Cu, that has exerted adverse effects on the surrounding ecosystems. The soils around the factory were strongly acidified with pH 3 - 4.9, and contained high concentrations of Pb and Zn (total Pb and Zn: over 1200 mg kg^{-1}, and the DTPA-extractable Pb and Zn: over 100 mg kg^{-1}), leading to the completely devoid of vegetation on the surrounding land and subsequent water erosion. Many attempts, from both the factory and academic institutions, were done to re-vegetate the contaminated area with over forty plant species (including trees, shrubs and grasses). Unfortunately, most of them failed due to the hostile soil and air conditions, and only several plants (include *Paulownia tomentosa*, *Leucaena glauca*, *Nerium indicum*, *Paederia scandens*, *Cynodon dactylon*) showed relatively high tolerance of the edaphic and atmospheric conditions. To improve the growth performance of such plants, about 50 cm of the top soil was mixed with pond sediment and complex NPK fertilizer in order to dilute heavy metal concentrations and improve soil conditions. The reclamation project was quite successful after two years of cultivation of mixed plant species. However, the growth performance was still poor at the severely eroded area with the total canopy cover of 30-50%, resulting in the failure of erosion control. Vetiver grass was introduced to the most eroded area in an effort to control the erosion. After five months of cultivation, Vetiver established well on the contaminated land with the total canopy cover (including

P. tomentosa and Vetiver) of about 80%. The results from the later inspection indicated that the erosion of the area planted with Vetiver was under control (Shu and Xia, 2003).

Chaiwat Phadermrod (2015) reported a highly successful large-scale rehabilitation of a zinc mine carried out by Padaeng Industry Public Company Limited (PDI) in Mae Sot District and its refinery is in Tak Province. The mine has grown vetiver for rehabilitation in the last 12 years concurrent with mining operation. A total of 19.17 million Vetiver slips were planted. Therefore, PDI mine is one of the biggest mines in Thailand where Vetiver has been grown to protect the environment. PDI mine is growing 1-2 million Vetiver slips every year, and planting trees at the same time (Figures 74-75).

Bare soil is planted with Vetiver first for soil rehabilitation, adding more organic matter into the soil, preventing soil erosion, reducing velocity of runoff water and protecting moisture in the soil. Then local tree species, such as teak, iron wood, Siamese sal, local cork tree, orchid tree and others were planted. From 1993 to 2014, an area of 166 ha (or 62% of leases) was rehabilitated at the cost of 63 million Thai Baht, the company will return the whole area with plantation forest to the Royal Forest Department. The company hopes that all stakeholders including the surrounding communities will protect the plantation forest area after post mining for their own benefit forever.

Figure 74. Padaeng mine pit.

Figure 75. Padaeng mine year 2003 and 2013.

6.7. Iron ore mine

Old and new overburdens together with waste dump at the Joda mine of Tata Steel in Bengal, India are very unstable and highly erodible due to extremely steep gradient (Figure 76). Due to the extremely adverse conditions, reduction of slope gradient to about 30°-40° by earth works and high quality planting materials were required for revegetation of the contaminated areas using Vetiver. The grass was planted at vertical interval between 1 and 1.5 m depending on slope gradients, and watered after planting every day. Excellent establishment and growth were achieved even on the slope of 40°, with root length reaching 60 cm on twenty day old plants. Once stabilised, the microenvironment created by vetiver hedges encouraged the return of native flora and even the opportunity for inter-cropping between hedges in a land shortage country like India (Figure 77) (K Pathak, pers.com).

Figure 76. The old waste dump is highly erodible and steep (top and bottom left), and re-shaped to 40° (bottom right).

Figure 77. Vetiver planted at vertical interval of 1-1.5 m (top), 6 months after planting and intercropping with vegetable (bottom).

6.8. Ammonia and nitrate contaminated land

The site, at Bajool, Australia, was contaminated with extremely high levels of ammonia and nitrate as a result of explosive manufacturing (Table 15). Vetiver grass was applied at this site as a phytoremediation measure to remove ammonia and nitrate and to stabilise the site, preventing offsite pollution by runoff with potential to contaminate the local environment.

One year after planting, the results were outstanding despite severe drought, excellent growth resumed following heavy rain six months later (Figure 78). With this high growth rate, this planting removed at least 629 kg ha^{-1} year^{-1}, and possibly much higher, at 1100 kg ha^{-1} year^{-1} depending on the rain. Hence it is projected that most of the N in the fill will be removed by

Vetiver in less than 4 years under favourable weather and at most 6 years under normal weather conditions.

Table 15. The characteristics of the contaminated site at Bajool, Australia.

Site features	Units	
Land surface area	m^2	7300
Soil depth	M	2.5 – 3
Contaminated soil volume	m^3	20000
Soil ammonia level	mg kg^{-1}	Range: 20 – 1220 Average: 620
Soil total nitrogen level	mg kg^{-1}	Range: 31 – 5380 Average: 2700
Water ammonia level	mg kg^{-1}	Range: 235 – 1150 Exceptional highest value: 12500
Water total nitrogen level	mg kg^{-1}	Range: 118 – 7590 Exceptional highest value: 18300

Figure 78. Vetiver planted at the contaminated site (top), 3 (bottom right) and 12 (bottom left) months after planting.

Another site at Peak Downs, Australia was contaminated with ammonia, nitrate as well as agro-chemicals that had been used for weed control around the mine. The contamination level at the Peak Downs varies greatly from very low to very high (soil ammonia: 20 – 3800 mg kg^{-1}, soil total nitrate: 10 – 7620 mg kg^{-1}) depending on the location of the site. Excellent growth of Vetiver was recorded as expected under good irrigation and high N level in the soil. To satisfy the P demand of Vetiver under this high N supply, superphosphate fertilizer was applied at planting.

The above results indicate that Vetiver grass can be established on soils contaminated with elevated levels of ammonia and nitrate. With appropriate management and maintenance, Vetiver can be effectively used for decontamination of these sites.

6.9. Hydrocarbon contaminated land

An oil shale dump, with an area of 667 ha and a depth of several to ten meters, is situated in the north suburb of Maoming City, Guangdong province, China. The oil shale dump was mainly composed of oil refined wastes with relatively high contents of organic matter, up to 3.61%, and mixed with infertile soils coming from oil shale excavation. The contents of N, P and K nutrients, especially their available contents, were distinctly scarce. The pH of soil and leachate was quite low, 4 and 3.2, respectively. Heavy metal concentrations in the soil were from the lowest 0.1 mg kg^{-1} for Cd to the highest 59.5 mg kg^{-1} for Mn. The combination of these adverse physical and chemical features yielded an environment hostile to organisms and, therefore, it was quite difficult to re-vegetate it (Xia, 2004).

A field plot trial was conducted to investigate the growth of Vetiver, Bahia, St. Augustine, and Bana in this oil shale dump amended with inorganic fertilizer and fishpond sludge. Results indicated that Vetiver had the highest survival rate, up to 99%, followed by Bahia and St. Augustine, 96% and 91%, respectively, while Bana had the lowest survival rate of 62%. The coverage and biomass of Vetiver were also the highest after 6-month planting. Fertilizer application significantly increased biomass and tiller number of the four grasses, of which St. Augustine was promoted most, up to 70% for biomass, while Vetiver was promoted least, only 27% for biomass. It can be concluded that Vetiver can be sustainably grown on this hydrocarbon infested tailings with minimal or without fertilizing (Xia, 2004).

6.10. Agricultural chemicals contaminated land

Castorina (2015) conducted a greenhouse experiment comparing the efficiency of Vetiver and canola (*Brassica napus* L.) for the phytoremediation of land in the "Valle del sacco" near Rome, Italy) where the natural soil environment had been altered by agricultural chemicals and improper disposal of industrial waste, causing a series of diseases in people and animals (Figure 79).

To assess the uptake of elements by the plants a total content analysis was done of the soil, followed by an analysis of the extractible fraction in EDTA (ethylene-diaminetetraacetic acid). The analytical data obtained were used to determine the

Translocation Factor (TF) and the Bio-concentration Factor (BF) of each toxic element for the two plants under the two different agricultural conditions, fertilised and unfertilised pots.

Analytical results of the soil show that the levels of As, Be, Cd, Co, Cr, Cu, Pb, V and Zn were significantly below the acceptable values for commercial and industrial sites. They are also below the values for public, private and residential green zones, except for lead, which is at the threshold limit.

The EDTA extractable fractions in soil were in many cases were much more significant, higher than 10% for Mo, Cu and Cd; and 20% for Pb, Co and Mn. After only a 5-month growth period, for many elements (e.g. Al, Cd, Cu, Fe, Pb and Zn) there was a significant decrease in the EDTA extractable fraction. In some cases (e.g. Ti and V), an increase was noted in the EDTA extractable fraction in the soil after the plants were extracted, and this was true for both canola and Vetiver. Moreover, for both plants, soil EC (Electric Conductivity) decreased after harvesting, by 50% for Vetiver. Phosphate fertilization increased the TF in both canola and Vetiver.

For many elements, Vetiver showed a higher BF than canola, but the TF was generally lower compared to canola. While the BF calculated with regard to the total elemental content are non-significant (very low values), those calculated with regard to the EDTA extractable fractions are more significant, especially for some elements such as Cr, Ti and Zn.

Figure 79: Contaminated site and Canola (*Brassica napus* L.)

REFERENCES

Adams R.P., Dafforn M.R. (1997). DNA fingertyping (RAPDS) of the pantropical grass vetiver (Vetiveria zizanioides L.) reveals a single clone "sunshine" is widely utilised for erosion control. The Vetiver Network Newsletter, no.18. Leesburg, Virginia USA.

Aldana E., Saffon I., Arcila, J., Ortiz M., Herrera O. (2013). Remoción de Aluminio en aguas residuales industrials usando especies macrófitas: Una aplicación para el pasto Vetiver. The second Latin America International Conference on the Vetiver System, October 3-5 2013, Medillin, Colombia (in Spanish).

Angin I., Turan M., Quirine M., Cakici, A. (2008). Humic acid addition enhances B and Pb phytoextraction by Vetiver grass (*Vetiveria zizanioides* (L.). Nash). Water Air Soil Pollut. , 188, 335-343

ANZ (1992). Australian and New Zealand Guidelines for the Assessment and Management of Contaminated Sites. Australian and New Zealand Environment and Conservation Council, and National Health and Medical Research Council, January 1992.

Arochas A., Volker K., Fonceca R. (2010). Application of of Vetiver grass for mine sites rehabilitation in Chile. Latin American Vetiver Conference, Santiago, Chile, Oct. 2010.

Ash R., Truong P. (2003). The use of Vetiver grass wetlands or sewerage treatment in Australia. The Third International Conference on Vetiver, Guangzhou, China, 6-9 October 2003.

Asokan P., Saxena M., Asolekar S.R. (2005). Coal combustion residues: environmental implications and recycling potentials. Resources, Conservation and Recycling, 43, 239-262

Barapanda P., Singh S.K., Pal B.K. (2001). Utilization of coal mining wastes: in Mining and Allied Industries, Regional Engg College, Rourkela, Orissa, India.

Berg J.V.D (2006). Vetiver grass (*Vetiveria zizanioides* (L.) Nash) as trap plant for *Chilo partellus* (Swinhoe) (Lepidoptera: Pyralidae) and *Busseola fusca* (Fuller) (Lepidoptera: Noctuidae), Annales de la Société entomologique de France (N.S.). International Journal of Entomology, 42, 449-454.

Bertea C.M., Camusso W. (2002). Anatomy, biochemistry and physilogy. In: Vetiveria. The Genus Vetiveria, pp 19-43 (Maffei M. ed) Taylor and Francis Publ., London and New York.

Bevan O., Truong P., Wilson M. (2000). The use of Vetiver grass for erosion and sediment control at the Australian bentonite mine in Miles, Queensland. The Fourth Innovative Conference, Australian Minerals and Energy Environment Foundation: On the threshold: Research into Practice, Brisbane, Australia, August 2000, 124- 128.

Brandt R., Merkl N., Schultze-Kraft R., Infante C., Broll G. (2006). Potential of Vetiver (*Vetiveria zizanioides* (L.) Nash) for phytoremediation of petroleum hydrocarbon-contaminated soils in Venezuela. International Journal of Phytoremediation, 8, 273-284.

Burton G.W., Hanna W.W. (1985). Bermuda grass. In: Heath ME, Garnes RF, Metcalfe DS (eds.), Forages. Iowa State Univ. Press, Ames, Iowa.

Castorina, B. (2015). Efficiency of vetiver for the phytoremediation of contaminated land in the "Valle del Sacco" (Rome). The sixth International Conference on Vetiver, Vietnam, Danang, May 3-5, 2015.

Chaiwat P. (2015), Vetiver for rehabilitation of padaeng zinc mine, Mae Sot district, Tak Province, Thailand. The sixth International Conference on Vetiver, Vietnam, Danang, May 3-5, 2015.

Cheng H., Yang X., Liu A., Fu H., Wan M. (2003). A study on the performance and mechanism of soil-reinforcement by herb root system. The Third International Conference on Vetiver, Guangzhou, China, 6-9 October 2003.

Chomchalow, N. (2006). Review and update of the Vetiver System R&D in Thailand. Regional Vetiver Conference, Cantho, Vietnam.

Cull R.H., Hunter H., Hunter M., Truong, P. (2000). Application of Vetiver grass technology in off-site pollution control. II. Tolerance of Vetiver grass towards high levels of herbicides under wetland conditions. The Second International Conference on Vetiver, Thailand, Phetchaburi, 18-22 January 2000.

Cuong D.C., Minh V.V., Truong P. (2015). Effects of sea water salinity on the growth of Vetiver grass (*Chrysopogon zizanioides* L.). The sixth International Conference on Vetiver, Vietnam, Danang, May 3-5, 2015.

Danh L.T., Truong P., Mammucari R., Foster N. (2012). Phytoredemdiation of soils contaminated with salinity, heavy metals, metalloids, and radioactive materials. In 'Phytotechnologies: Remediation of Environmental Contaminants', edited by Naser A. Anjum, published by CRC Press/Taylor and Francis Group, Boca Raton, Florida, USA, pp 255-282.

Danh L.T., Truong P., Mammucari R., Tran T., Foster N. (2009). Vetiver grass, *Vetiveria zizanioides*: A choice plant for phytoremediation of heavy metals and organic wastes. International Journal of Phytoremediation 11, 664- 691.

Danh L.T., Phong L.T., Dung L.V., Truong P. (2006). Wastewater treatment at a seafood processing factory in the Mekong delta, Vietnam. The Fourth International Conference on Vetiver, Caracas, Venezuela, October 2006.

Darajeh N., Idris A., Truong P., Aziz A.A., Bakar R.A., Man H.C. (2014). Phytoremediation potential of Vetiver System Technology for improving the quality of palm oil mill effluent. Advances in Materials Science and Engineering, Volume 2014, Article ID 683579, 10 pages.

Das P., Datta R., Makris K.C., Sarkar D. (2010). Vetiver grass is capable of removing TNT from soil in the presence of urea. Environmental Pollution, 158, 1980–1983.

Das M., Adholeya A. (2009). A short comparison study on growth of *Vetiver zizanioides* with different AM species on fly ash. Mycorrhiza News, 21, 19-24.

Datta R., Das P., Smith S., Punamiy P., Ramanathan D.M., Reddy R, Sarkar D. (2013). Phytoremediation potential of Vetiver grass [*Chrysopogon zizanioides* (L.)] for tetracycline. International Journal of Phytoremediation, 15, 343–351.

Fonseca R., Diaz C., Castillo M., Candia J., Truong P. (2006). Preliminary results of pilot studies on the use of Vetiver grass for mine rehabilitation in Chile. The Fourth International Conference on Vetiver, Caracas, Venezuela, October 2006.

Ghosh M, Paul J., Jana A., De A., Mukherjee A. (2015). Use of the grass, *Vetiveria zizanioides* (L.) Nash for detoxification and phytoremediation of soils contaminated with fly ash from thermal power plants. Ecological Engineering, 74, 258–265.

Greenfield J.C. (2002). Vetiver Grass: An essential grass for conservation of planet earth. Infinity Publishing Co, Haverford, PA, USA.

Hart B., Cody R., Truong P. (2003). Efficacy of Vetiver grass in the hydroponic treatment of post septic tank effluent. The Third International Conference on Vetiver, Guangzhou, China, 6-9 October 2003.

Hatch M.D. (1987). C4 photosynthesis: a unique blend of modified biochemistry, anatomy and ultrastructure. Biochimica et Biophysica Acta. 895, 81-106.

Hengchaovanich D. (1998). Vetiver grass for slope stabilization and erosion control, with particular reference to engineering applications. Pacific Rim Vetiver Network Technical Bulletin 2.

Hengchaovanich D., Nilaweera N.S. (1998). An assessment of strength properties of Vetiver grass roots in relation to slope stabilization. The First International Conference on Vetiver, Chiang Rai, Thailand, 4-8 February 1998.

Hengchaovanich D. (1999). Fifteen years of bioengineering in the wet tropics from A (*Acacia auriculiformis*) to V (*Vetiveria zizanioides*). The First Asia-Pacific Conference on Ground and Water Bio-engineering, Manila, Philippines, April 1999.

Hung L.V., Cam B.D., Nhan D.D., Van T.T. (2012). The uptake of uranium from soil by vetiver grass (*Vetiver zizanioides* (l.) Nash) Vietnam Journal of Chemistry, 50, 656-662.

Jala S., Goyal D. (2006). Fly ash as a soil améliorant for improving crop production: a review Bioresource Technology, 97, 1136-1147.

Jalalipour H., Haghighi A.B., Truong P. (2015). Vetiver phytoremediation technology for rehabilitating Shiraz municipal landfill, Iran. The 6th International Conference onVetiver, Vietnam, Danang, 3-5 May, 2015.

Inman-Bamber N.G. (1974). CANEGRO, its history, conceptual basis, present and future uses. Workshop on Research and Modeling Approaches to Examine Sugarcane Production Opportunities and Constraints, St Lucia, Queensland.

Kingett Mitchell and Associates (1995). An assessment of urban and industrial stormwater inflow to the Manukau Harbour, Auckland. Regional Waterboard Techn. Publ. No. 74.

Lavania S. (2003). Vetiver Root System: Search for the Ideotype. The Third International Conference on Vetiver, Guangzhou, China, 6-9 October 2003.

Le Viet Dung (2015). Report on the Research, Development and Promotion of the Vetiver System at Cantho University, Vietnam from 2002 to 2012. Cantho University Publication (In Vietnamese).

Leaungvutiviroj C, Piriyaprin S, Limtong P, Sasakic K. (2010). Relationships between soil microorganisms and nutrient contents of *Vetiveria zizanioides* (L.) Nash and *Vetiveria nemoralis* (A.) Camus in some problem soils from Thailand. Applied Soil Ecology 46: 95-102.

Lee O. (2013). The Vetiver latrine. The Second Latin American Vetiver Conference, Medellin, Colombia, 3-5 October 2013.

Li H., Luo Y.M., Song J., Wu L.H., Christie P. (2006). Degradation of benzo[a]pyrene in an experimentally contaminated paddy soil by Vetiver grass (*Vetiveria zizanioides*). Environmental Geochemistry and Health, 28, 183–188.

Liao X., Shiming, L., Yinbao, W., Zhisan, W. (2003). Studies on the abilities of *Vetiveria zizanioides* and *Cyperus alternifolius* for pig farm wastewater treatment. The Third International Conference on Vetiver, Guangzhou, China, 6-9 October 2003.

Lomonte C., Wang Y., Doronila A., Gregory D., Baker A.J.M., Siegele R., Kolev S.D. (2014). Study of the spatial distribution of mercury in roots of Vetiver grass (*Chrysopogon zizanioides*) by micro-pixe spectrometry. International Journal of Phytoremediation,16, 1170-1182.

Luque R., Lisena M., Luque O. (2006) Vetiver System For Environmental Protection of Open Cut Bauxite Mining At "Los Pijiguaos" –Venezuela. The Fourth International Conference on Vetiver, Caracas, Venezuela, October 2006.

Makris K.C., Shakya K.M., Datta R., Sarkar D., Pachanoor D. (2007a). High uptake of 2,4,6-trinitrotoluene by Vetiver grass - Potential for phytoremediation ? Environmental Pollution, 146, 1-4.

Makris K.C., Shakya K.M., Datta R., Sarkar D., Pachanoor D. (2007b). Chemically catalyzed uptake of 2,4,6-trinitrotoluene by *Vetiveria zizanioides*. Environmental Pollution, 148, 101-106.

Marcacci S., Schwitzguébel J.P., Raveton M., Ravanel P. (2006). Conjugation of atrazine in vetiver (*Chrydopogon zizanioides* Nash) grown in hydroponics. Environmental and Experimental Botany. 56: 205 - 215.

Materechera S. (2010). Soil physical and biological properties as influenced by growth of Vetiver grass (*Vetiveria zizanioides* L.) in a semi-arid environment of South Africa. 19th World Congress of Soil Science, Soil Solutions for a Changing World 1 – 6 August 2010, Brisbane, Australia.

Mickovski S.B., van Beek L.P.H., Salin F. (2005). Uprooting of Vetiver uprooting resistance of Vetiver grass (*Vetiveria zizanioides*). Plant and Soil, 278, 33–41.

Monteiro J.M., Vollú R.E., Coelho M.R.R., Alviano C.S., Blank A.F., Seldin L. (2009). Comparison of the bacterial community and characterization of plant growth-promoting rhizobacteria from different genotypes of *Chrysopogon zizanioides* (L.) Roberty (Vetiver) rhizospheres. The Journal of Microbiology 47: 363-370.

Muchow R.C., Sinclair T.R., Bennett J.M. (1990). Temperature and solar radiation effects on potential maize yield across locations. Agronomy Journal, 82, 338-343.

National Resource Council, 1995. Vetiver Grass: a thin green line against erosion. National Academy Press, Washington, DC.

Huong N.T.T., Van, T.T., Truong, P. (2015). Effectiveness of Vetiver grass in phytostabilization and/or phytoremediation of dioxin-contaminated soil at Bien Hoa airbase, Vietnam. The 6th International Conference on Vetiver, Vietnam, Danang, 3-5 May, 2015.

Nix K.E., Henderson G., Zhu B.C.R., Laine R.A. (2006). Evaluation of Vetiver grass root growth, oil distribution, and repellency against Formosan subterranean termites. Hort. Science, 41, 167-171.

Noffke R. (2013). Mine and associated rehabilitation projects in Africa and Indian ocean islands. The Second Latin American Vetiver Conference, Medellin, Colombia, 3-5 October 2013.

Northcote K.H., Skene J.K.M. (1972). Australian Soils with Saline and Sodic Properties. CSIRO Div. Soil. Pub. 27.

Oku E., Asubonteng K., Nnamani C., Michael I., Truong P. (2015). Using native African species to solve African wastewater challenges: An in-depth study of two Vetiver grass species. The Sixth International Conference on Vetiver, Vietnam, Danang, 3-5 May, 2015.

Percy I., Truong P. (2005). Landfill leachate disposal with irrigated Vetiver grass. National Conference on Landfill, Brisbane, Australia, Sept 2005

Percy I., Truong, P. (2003). Landfill leachate disposal with irrigated Vetiver grass. The Third International Conference on Vetiver, Guangzhou, China, 6-9 October 2003.

Phenrat T., Teeratitayangkul P., Imthiang T., Sawasdee Y., Wichai S., Piangpia T., Naowaopas J., Supanpaiboon W. (2015). Laboratory-scaled developments and field-scaled implementations of using Vetiver grass to remediate water and soil contaminated with phenol and other hazardous substances from illegal dumping at Nong Nea subdistrict, Phanom Sarakham district, Chachoengsao province, Thailand. The Sixth International Conference on Vetiver, Vietnam, Danang, 3-5 May, 2015.

Rai A.K., Paul B., Singh G. (2011). A study on physico chemical properties of overburden dump materials from selected coal mining areas of Jharia coalfields, Jharkhand, India. International Journal of Environmental Sciences, 1, 1351-1360.

Radloff B., Walsh K., Melzer A. (1995). Direct Revegetation of Coal Tailings at BHP. Saraji Mine. Australian Mining Council Environmental Workshop, Darwin, Australia.

Roongtanakiat N., Osotsapar Y., Yindiram C. (2008). Effects of soil amendment on growth and heavy metals content in Vetiver grown on iron ore tailings. Kasetsart J. (Nat. Sci.), 42, 397-406.

Ruiz C., Rodríguez O., Luque O., Alarcón M. (2013). Effecto del vetiver (Chrysopogon zizanioides L.) en la reducción del flúor y otros compuestos contaminantes en aguas de consumo humano. Caso: Caserío Guarataro, estado Yaracuy, Venezuela. The second Latin America International Conference on the Vetiver System, October 3-5 2013, Medillin, Colombia.

Shu W, Xia H. (2003). Integrated Vetiver Technique for Remediation of Heavy Metal Contamination: Potential and Practice. The Third International Conference on Vetiver, Guangzhou, China, 6-9 October 2003.

Shu W.S., Zhao Y.L., Yang B., Xia H.P., Lan C.Y. (2004). Accumulation of heavy metals in four grasses grown on lead and zinc mine tailings. Journal of Environmental Science, 16, 730-434.

Singh S., Melo J.S., Eapen S., D'Souza S.F. (2008). Potential of Vetiver (Vetiveria zizanoides L. Nash) for phytoremediation of phenol. Ecotoxicology and Environmental Safety 71 (2008) 671–676.

Siripin S., Thirathorn A., Pintarak A., Aibcharoen P. (2000). Effect of associative nitrogen fixing bacterial inoculation on growth of Vetiver grass. The Second International Conference on Vetiver, Phetchaburi, Thailand, 18-22 January 2000.

Smeal C., Hackett M., Truong P. (2003). Vetiver System for industrial wastewater treatment in Queensland, Australia. The Third International Conference on Vetiver, Guangzhou, China, 6-9 October 2003.

Thao Minh Tran, Lacoursière, J.O., Vought, L.B.M., Phuong Thanh Doan and Man Van Tran. (2015). Capacity of vetiver grass in treatment of a mixture of laboratory and domestic wastewaters. The Sixth International Conference on Vetiver, Vietnam, Danang, 3-5 May, 2015.

Triana R., Burgos J., Zúñiga J. (2013). Piloto de tratamiento no convencional para aguas asociadas a la producción de hidrocarburos empleando humedal artificial con pasto vetiver. The Second Latin American Vetiver Conference, Medellin, Colombia, 3-5 October.

Truong P., Creighton C. (1994). Report on the potential weed problem of Vetiver grass and its effectiveness in soil erosion control in Fiji. Division of Land Management, Queensland Department of Primary Industry, Brisbane, Australia.

Truong P., Baker D. (1997). The role of Vetiver grass in the rehabilitation of toxic and contaminated lands in Australia. International Vetiver Workshop (), Fuzhou, China, 21-26 Octocber.

Truong P. (1999). Vetiver Grass Technology for mine rehabilitation. In: Ground and water bioengineering for erosion control and slope stabilisation, pp 379-389 (Barker, D. H. et al. ed) Sciences Publishers, New Hampshire USA.

Truong P.N., Hart, B. (2001). Vetiver system for wastewater treatment. Technical Bulletin No. 2001/2. Pacific Rim Vetiver Network. Office of the Royal Development Projects Board, Bangkok, Thailand.

Truong P. (2002). Vetiver Grass Technology. In: Vetiveria. The genus Vetiveria, pp 114-132 (Maffei M. ed) Taylor and Francis Publ., London and New York.

Truong P. (2003). Vetiver System for water quality improvement. The Third International Conference on Vetiver, Guangzhou, China, 6-9 October.

Truong P., Smeal C. (2003). Research, Development and Implementation of Vetiver System for Wastewater Treatment: GELITA Australia. Technical Bulletin No. 2003/3. Pacific Rim Vetiver Network. Office of the Royal Development Projects Board, Bangkok, Thailand

Truong P., Truong S., Smeal, C (2003). Application of the Vetiver System in Computer Modelling for Industrial Wastewater Disposal. The Third International Conference on Vetiver, Guangzhou, China, 6-9 October 2003

Truong P., Van T.T., Pinners E. (2008). Vetiver System Applications: A Technical Reference Manual. The Vetiver Network International.

Truong P., Booth D. (2010). Final Report on the Application of Vetiver System in the Citarum River Basin, Indonesia. The Indonesian Vetiver Network.

Truong P.N., Granley B.A., Calderon M. (2012). Leachate treatment with phytoremediation: Case Studies. Global Waste Management Symposium. Phoenix, Arizona, USA, September 2012.

Truong P., Truong N. (2013). Computer model for treatment of small volume waste water. Proc. Second Latin America International System Conference on Vetiver, Medellin, Colombia, 3-5 October.

Ugalde Smolcz, S. and Goykoviv Cortés, V. (2015). Remediation of boron contaminated water and soil with Vetiver phytoremediation technology in northern Chile. The sixth International Conference on Vetiver, Vietnam, Danang, May 3-5, 2015.

Van T.T., Truong P. (2008). R&D results on unique contributors of Vetiver applicable for its use in disaster mitigation purposes in Vietnam. The First Indian National Vetiver Workshop, Cochin, India, 21-23 February 2008.

Vieritz A., Truong P., Gardner T., Smeal, C (2003). Modelling Monto Vetiver growth and nutrient uptake for effluent irrigation schemes. The Third International Conference on Vetiver, Guangzhou, China, 6-9 October 2003.

Vose J.M., Harvey G.J., Elliott K.J., Clinton B.D. (2004). Measuring and Modeling Tree and Stand Level Transpiration. In Phytoremediation: Transformation and Control of Contaminant. John Wiley & Sons, Inc.

Wagner S., Truong P., Vieritz A. (2003). Response of Vetiver grass to extreme nitrogen and phosphorus supply. The Third International Conference on Vetiver, Guangzhou, China, 6-9 October 2003.

Wang Y.W. (2000). The root extension rate of Vetiver under different temperature treatments. The Second International Conference on Vetiver, Phetchaburi, Thailand, 18-22 January 2000.

Winter S. (1999). Plants reduce atrazine levels in wetlands. Final year report. School of Land and Food, University of Queensland, Brisbane, Queensland, Australia.

Xia H. P, Ao H. X, Lui S, H. and He D. Q. (1997). A premilitery study on vetiver's purification for garbage leachate. International Vetiver Grass Workshop. Fuzhou. China. http:// www.vetiver.org

Xia H.P. (2004). Ecological rehabilitation and phytoremediation with four grasses in oil shale mined land. Chemosphere, 54, 345–353.

Xia H.P, Lu X., Ao H., Liu S. (2003). A preliminary Report on Tolerance of Vetiver to Submergence. The Third International Conference on Vetiver, Guangzhou, China, 6-9 October 2003.

Xia H.P., Ao H.X., Lui S.Z., He D.Q. (1999). Application of the Vetiver grass bioengineering technology for the prevention of highway slippage in Southern China. Proceeding of Ground and Water Bioengineering for Erosion Control and Slope Stabilisation, Manila, April 1999.

Yang B., Shu W.S., Ye Z.H., Lan C.Y., Wong M.H. (2003). Growth and metal accumulation in Vetiver and two Sesbania species on lead/zinc mine tailings. Chemosphere, 52, 1593–1600.

Zhang X., Gao B., Xia H. (2014). Effect of cadmium on growth, photosynthesis, mineral nutrition and metal accumulation of bana grass and vetiver grass. Ecotoxicology and Environmental Safety, 106, 102–108.

Zheng C.R., Tu C., Chen H.M. (1997). Preliminary study on purification of eutrophic water with Vetiver. International Vetiver Workshop, Fuzhou, China.

www.ingramcontent.com/pod-product-compliance
Lightning Source LLC
Chambersburg PA
CBHW051018180526
45172CB00002B/398